THE
COMPLETE BOOK
OF THE
HORSE

THE COMPLETE BOOK OF THE HORSE

GENERAL EDITOR
Carol Foster

Colour Library Books

This edition published in 1991 by
Colour Library Books Ltd
Godalming Business Centre
Catteshall lane
Godalming
Surrey GU7 1XW
Telephone 0483 426266

by arrangement with The Hamlyn Publishing Group
Michelin House
81 Fulham Road
London SW3 6RB
© Octopus Books 1983

ISBN 0 86283 866 5

Printed by Mandarin Offset in Hong Kong

CONTENTS

The Horse and its Management

Popular Breeds of the World

The Competitive Spirit

FOREWORD
Bruce Davidson

As THE YEARS go by horses remain the main interest in my life, and I so much enjoy observing the pleasures, the confrontations, the many mixed attachments they promote and endure for so many different types of people.

If one is getting one's very first pony there are books and pamphlets galore on 'How To', or if you are about to embark on your first fox hunt one can read chapters or sections on just what to wear or the best ways of avoiding embarrassment. If you happen to have the friends that have just the right book with that particular chapter, or if you have the time and opportunity to research your subject.

I admire the tremendous commitment assembling a book of this complexity calls for and applaud its long awaited appearance. One thing is certain, I have read my synopsis on Eventing, yet never before was the subject so logically presented. It not only tells one precisely what is required in each discipline but goes into the detail and explanation that only the experience of years of fine horsemanship can produce and compiled to interest horsemen of any background.

If one really loves horses, one loves every aspect of horses, be it polo, hunting, racing, stockseat, eventing etc. There is something to be learned from every aspect, and *The Complete Book of The Horse* gives one a good experience in the horse world.

Horses are much more than just a childish fancy or an extremely stylish thing to do. Whatever your connection, new or old, perhaps you have that incurable disease, horse-fever, that so many horse lovers contract. To those who have dedicated their lives to the equine world, in whatever roll you fill, be it owner, groom, rider, parent, or dreamer, dreaming of what it will be like when you get the chance to have a close connection with a horse, let one of your first experiences be with this book! I found the text to be helpful for brief reference to forgotten detail as well as enjoyable and instructional for those who wish to expand their knowledge and understanding of the horse and its human connection.

Any subject if followed and studied deeply enough has aspects and sections of complexity most people never consider. If every horse lover had access to such knowledge on so many, varied horse subjects the world would be a much better place for horses. *The Complete Book of the Horse* does a terrific job in improving the odds that horses will be better understood and looked after in every respect, and one hardly need remind the already knowledgeable horseman that there is always more to learn, for surely they will have a copy.

PREFACE
Carol Foster

THERE IS MORE to owning a horse than feeding, watering and riding him. He is your leisure and enjoyment, but he is also a living creature whose welfare should be taken seriously. Just as it is sensible to know at least the rudiments of car mechanics so, to gain the greatest fulfilment from your horse, and to care and ride him to the best of your ability, you should learn about his 'mechanics', how he lives and works and later, why he behaves in the way he does and the influence your own mind and body have upon him.

This book is for you and your horse. Learning to ride is but the beginning of years of fun in prospect with your own horses who will teach you not only that you will never stop learning about them and the sport they provide, but also more about yourself, physically and mentally. No book, therefore, can be a substitute for experience itself but it is easy for experience to pass by without an awareness of opportunites open to you and your horse.

On an international scale people are taking to the horse, learning to ride and, in increasing numbers, buying their own horses. Advice on buying your horse, and the pitfalls to be avoided, may be read in conjunction with the extensive section on light horse breeds which indicates the purposes to which each breed or type is particularly suited.

The veterinary chapter tells you not just what may go wrong but, through an introduction to the biology of the horse, gives a clearer idea of why things go wrong and, by understanding the roots of problems, treatment and care become logical when prescribed by the veterinary surgeon. The care and management chapters will not teach you to be an expert – again, only experience will teach you that – but they will guide you in a basic approach to what is best for the horse, the horsemaster's primary aim. Time and experience, sometimes bitter, will make you wiser providing you have learnt from the experiences and made the most of the observing and questioning the experts.

The subject is vast and may seem bewildering, if not off-putting, but the authors who will arouse the competitive spirit in you are all experts who 'do' as well as write, and their aim is to steer you onto the road they have taken and give you guidelines for your progress. Western Riding is a law unto itself but the rapport western riders have with their horses, and the innate intelligence of the horses themselves is a lesson for all riders. The art of the coachman, almost lost, has made an exciting return to competitive performance trials – the three-day event of the driving world while, side-saddle riding, showing and hunting, far from anachronisms, gain in popularity and, like driving, retain the elegance of a former age.

The ancient and classical art of dressage attracts more devotees annually while for the brave and young at heart the speed, thrills and accuracy demanded by show jumping and, more, eventing, have no peers.

Riding has never enjoyed so much popularity and the success and growth of international competitions is but the tip of the iceberg. For every top international 'name', there are hundreds of riders on the slippery slope (for there are always setbacks) and even more who still haven't surfaced: two-thirds of an iceberg is underwater. Very few people do not have a competitive streak, even if competing only against themselves in a desire for exhilaration through self-discipline and your horse opens the door for you to take part in one or more of the wide range of equestrian activities. The international aspect means that similar competitions are held the world over, under much the same rules – the horse unites all nationalities in a mutual understanding.

The base of that 'iceberg' is very broad and opportunity abounds for everyone, whether you want merely to have a good time or to combine that with the extra determination which will lead to success in competitions. You don't have to be a professional to succeed and this book will help you along that path, whether it creates just an awareness of the many facets of the sport or gives you concrete advice as you go on your way.

The Horse and its Management

ADVICE ON BUYING your horse, and the pitfalls to be avoided, starts this section of the encyclopedia. It then moves onto horse care and management providing valuable information on these two subjects. This second chapter will guide you in your approach as it is written by an author with a wealth of experience in this area and should make you a wiser horsemaster. The final chapter in this section deals with veterinary care, it tells you not just what may go wrong, but gives a clearer idea of why things do go wrong.

1

BUYING A HORSE

Whether or not you already have your own horse, sooner or later, cheque book at the ready, you will find yourself in the market for a new or replacement model. There cannot be many people in the horse world who have come through their dealings completely unscathed, so, if you're in that unfortunate position at the moment, these notes may help you in your next selection. And if you're eagerly anticipating buying your first horse, put your cheque book away for a few minutes and read on!

What do you want?

There are many considerations to be made and decisions to be reached before you take the big step of buying a horse. The first point you must establish is, why do you want a horse and what do you intend to do with it? Do you, for example, want to ride purely for pleasure or do you aim to take part in show jumping, cross-country events and other such competitions. Whatever your ambitions, you must be critical of yourself and, though you should look for a horse with which you can progress, there is little point in buying something clearly beyond your capabilities or powers of control. Hopefully you will long ago have come to the conclusion that you will never stop learning, and your professional instructor will be only too pleased to give advice on the sort of horse for which you ought to be looking, or indeed to accompany you when you view the prospective purchase.

Time and money

When you have decided what you want to do with your horse, work out just how much time you will have to spend on your riding. Many readers of this book will have to fit in a full-time job with their riding and, make no bones about it, if you want to improve, sacrifices of time and money will have to be made. Many is the rider who rises at 5.30 every morning to ride before going to work, and still finds there is precious little time. So the degree of schooling that your new horse already has may be determined for you, not only by your riding

experience but also by the time you will have to devote to it. A young horse may sound a very attractive proposition, but it requires unlimited time and, unless you are prepared to keep it on a schooling livery basis, with an experienced professional doing much of the work for you, you are unlikely to make much headway if your riding is restricted to an hour a day. If your time and expertise are limited, therefore, look for a horse which has progressed a stage or two further than you, so that it will help you learn and give you confidence. And if you really lack confidence, look for a 'schoolmaster' on which you can have some real fun as well as improving your riding.

Your next consideration also has to do with the amount of time—and money—available. How and where will you keep your horse? Some useful suggestions appear in the stable management chapter of this book, but if you are to stable your horse at home or at a DIY livery, you can allow a *minimum* of three hours a day for all your stable duties plus exercise. Even if the horse is hardy enough to live out, you will need to attend to him each day: and grassland has to be managed too. Is there a farrier locally, and will he come to you or will you have to take your horse to him?

To be realistic, if you want to own a horse and have to work, then a good livery is the ideal compromise and, although the outlay may seem great, the service you receive, if you've done your homework, is really very good value. After all, if you're buying your horse to enjoy as a hobby, you don't want to spend 75 per cent of the time worrying about him. Do consider the expense carefully, however, and don't forget that on top of your livery bill will be shoeing and veterinary fees, although the latter can be covered by insurance. Many riding schools are sympathetic towards the problems of working people and offer 'working' livery schemes under which they have the use of your horse for a specified number of hours each week and keep him for you for a reduced fee. Look at such schemes carefully though, so that you know what you are getting for your money.

Where to find your horse

Having decided what you want a horse for, and how much time and money you have to spend on it, where do you look? Generally, unless you are very knowledgeable yourself or have the benefit of an experienced adviser, it is better to avoid the sale ring where it is easy to buy an attractive or forlorn-looking animal on impulse. There may not be an opportunity to ride the horse or have him vetted, so, although you could find a real bargain, you could also buy a lot of trouble. The exception would be specialized sales, possibly held under the auspices of a breed society, or a sale of competition horses whose histories could be checked. Attractive though they may sound, racehorses sold out of training are usually not a good buy unless you really know what you're about. They usually need re-breaking and rarely make suitable riding horses.

Many people are reluctant to answer advertisements in the equestrian press, but genuine horses are sold in this way and this is where one can probably see the widest selection available. Studying advertisement columns can help give an idea of the price brackets into which various types of horses fall, but advertisement copy should always be read sceptically! If you're still keen once you've pulled the advertisement to bits, contact the seller, but question him intelligently, open everything he says to interpretation and assess whether it's worth actually going to see the horse. It's no use being told glibly that the horse has done this event or that show jumping—find out when and where. Ask for specific details and perhaps do some investigative work by using contacts in the area who may know the sellers of the horse.

Horse dealers have a much better name nowadays than they did in the past, and a reputable one in your area would probably agree to you taking a suitable horse on trial—providing, of course, you can meet his criteria for keeping the horse while in

Overleaf *The German Trakehner has the stamp of a good riding horse.*

Below left *A bold outlook over the stable door – your first meeting with a horse will tell volumes. Alert, pricked ears and a large generous eye indicate good character as well as looks.*

Right *A well-kept and fenced grass paddock with healthy horses should give the buyer confidence.*

Below right *See the horse ridden and put through all paces before you assess how it performs for you.*

your care. But if you do go to a dealer, do so by recommendation from someone who uses him regularly, such as your riding school.

Recommendation is probably the best way to find a horse, and once you are in the market and know what you're looking for, let as many people as possible know. Your 'scouts' are sure to come up with something sooner or later and it will be a horse whose history and temperament are known. Many riding establishments keep lists of horses for sale, acting as unofficial agents, so it is worth asking. Indeed, wherever you are—at a show or other competition—if you see a horse that catches your eye, wait for an opportune moment and make enquiries of the owners themselves. There are few horses that are not 'for sale' and you can lose nothing by making tentative enquiries. Don't expect to make money on your purchase, however. The 'average' rider will usually spend much more money than he ever recoups!

The main points of
the horse.

1 Poll
2 Forelock
3 Temple
4 Forehead
5 Haw (3rd eyelid)
6 Nose
7 Cheek
8 Muzzle
9 Chin
10 Chin groove
11 Throat
12 Jugular groove
13 Shoulder
14 Point of shoulder
15 Breast
16 Elbow
17 Forearm
18 Chestnut
19 Knee
20 Flexor tendons
21 Suspensory ligament
22 Cannon
23 Ergot
24 Fetlock
25 Pastern
26 Coronet
27 Hoof
28 Hoof wall
29 Heel
30 Crest
31 Neck
32 Withers
33 Back
34 Loins
35 Point of hip
36 Croup
37 Dock
38 Hip joint
39 Hindquarters
40 Buttock
41 Thigh
42 Hamstring (Achilles
   tendon)
43 Second thigh (Gaskin)
44 Shin
45 Point of hock
46 Hock
47 Back tendons
48 Stifle
49 Sheath
50 Flank
51 Belly
52 Ribs
53 Chest
54 Brisket

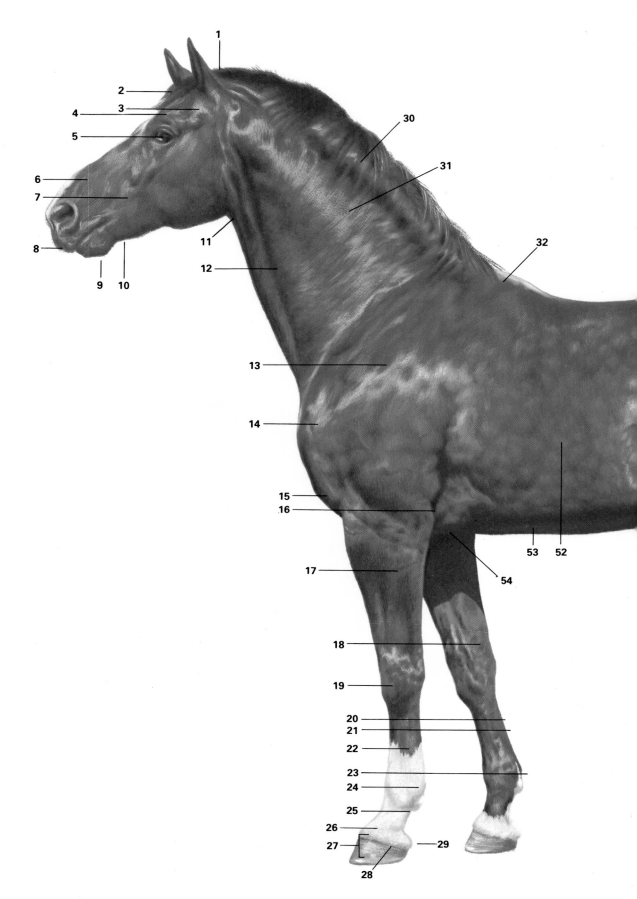

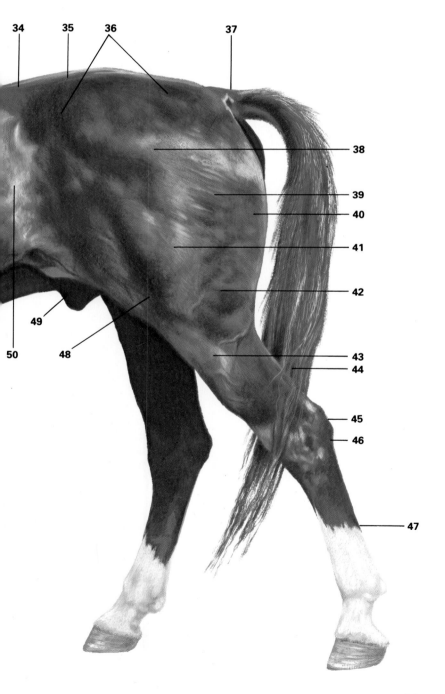

34 35 36 37
38
39
40
41
42
49
50 48
43
44
45
46
47

How to look at a horse

Whichever route brings you into first contact with your prospective purchase, the procedure which follows is common to all. Observance, attentiveness and a fair amount of perception are required; allow nothing to go unquestioned. An old nagsman's saying is 'Always look at the muck heap; never buy a horse from a yard where it isn't squared-up!', the state of the muck heap being indicative of the standard of the yard in general. Whether or not you decide to follow the advice to the letter is up to you, but you can gather a lot by the appearance of an establishment. Don't let anyone pull the wool over your eyes though, and however experienced you are, take someone with you for a second opinion.

You may already have formed an idea of the type of horse for which you are looking, depending again on your ambitions, but when you first inspect the horse, assess it with a critical eye. Of course, you can only generalize over the suitabilities of 'types' for a particular purpose—there are no hard and fast rules and there are many exceptions, but you have to start somewhere.

Before you see the horse led out of the box make a note of how he looks over his door. He should be taking an alert interest in the approaching party of humans, with ears pricked and maybe a call for the person he recognizes. Make sure you are given a chance to see him in the box and note how he reacts to people walking into the stable. He should remain amenable and show no objection to being handled. If things look good, pat him and stroke him; quietly put a hand up to his head and then run it down and over the back to the quarters; and give him another pat. Perhaps ask him to move over in the box. All the time watch the horse and gauge his stable manners. Ask if he is good to travel and, preferably, see him loaded into a trailer or lorry.

Make and shape

Ask the handler to lead the horse out of his box and note the horse's attitude. Does he walk out smartly and keenly, or is he lazy and reluctant? Is the handler using a headcollar or a bridle; if it is a bridle, ask if there is any particular reason—does the horse become 'bargey' when being led? Ask the handler to stand the horse up and assess his conformation. Does he look in proportion and balanced? While you may not worry about an attractive head, does the horse look honest with a generous eye and is the head set square onto the neck? If you want to do fast work, you don't want a horse with a stuffy throat and neck, for you want him to be able to breathe efficiently at speed. Never lose sight of the purpose for which you want the animal (reference to the breed and competition sections of this book will give you specific guidelines). Look for a good, sloping shoulder which will govern the smoothness of the horse's action and the ride he gives, and look at the horse from the front; you don't want a weak, 'pigeon chested' horse. Is he standing square? Look at his legs, and feel them—are they straight with flat knees and good bone (the measurement taken around the leg at the widest point below the knee). Do the legs look firm and 'clean' below the knee; there should be no puffiness or swelling and the tendons should feel hard and strong. If you feel a splint, try to assess whether it interferes with the knee action. The foot should be neither too flat nor too square and should be attached to the leg by a similar angle (neither too sloping nor too upright) through the pastern. Pick up the feet and examine them; is the horn hard and the frog well defined?

Move back along the horse. Look at his 'top line' and assess whether the neck moulds naturally into the back through well-defined withers. A horse which has a naturally rounded 'correct' outline gives an easier ride than one with a naturally 'hollow' outline. Assess the depth, or heart room, between the point of the withers and the girth and look for well-sprung but not 'barrelly' ribs. How long is the horse's back? The shorter the stronger, but the more compact the horse the less likely he is to be fast. If you want speed, you will want to see a horse which gives the overall picture of range and scope—more of a racehorse.

Move on to the 'powerhouse', the horse's quarters. Are they round and well muscled and is the tail set on well? Look for the vertical line from buttock to hock and fetlock which denotes strength, and again look for the well set-on pastern and foot. Feel the leg and pick up the foot, noting any possible problems, particularly any enlargements of the hock. Pay attention to any scars, especially on the legs, and ask what caused them.

The vet will give an accurate assessment of the horse's age when he carries out his inspection, but, even if you are not an expert, you can get some idea by looking at the teeth for wear and slope. The more mature the mouth, the more worn and discoloured the grinding surfaces of the teeth will appear. The teeth themselves will appear longer and project at a greater angle with age. The diagrams will give you some idea of what to look for at various ages.

Time for action

Now see the horse trotted up in hand, but watch the way it is done and ask the handler to run well away from the horse's head; if he seems to be pulling the horse towards him all the time you won't get an accurate impression of how the horse moves. Ask to see him walked away, then trotted towards and

Below 'Bone', which describes the build of the horse, is measured around the leg at the widest point below the knee. The greater the measurement the more weight the horse is capable of carrying.

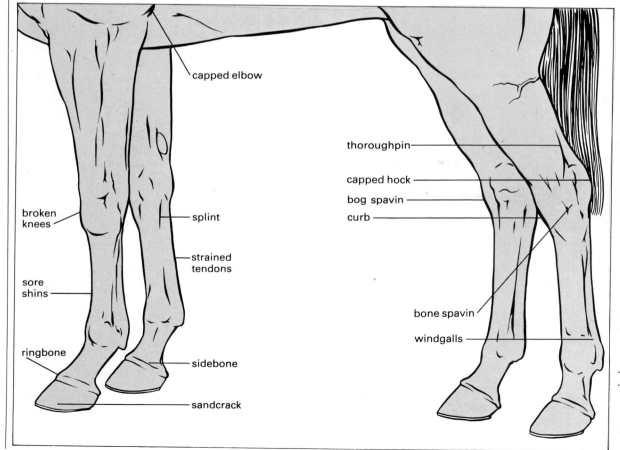

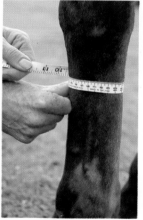

Left The seats of lameness. Lameness can result from bony growths such as splints, bone spavin and ringbone, swellings such as bog spavin, thoroughpin and windgalls, sandcrack and other hoof injuries, or from objects lodged in the foot. Watch for signs of limping as the horse is trotted out and feel the legs for injuries or swellings.

capped elbow

broken knees

splint

sore shins

strained tendons

ringbone

sidebone

sandcrack

thoroughpin

capped hock

bog spavin

curb

bone spavin

windgalls

past you, fast and freely. You will then be able to assess whether the horse is sound and straight. You don't want front legs that dish or hind legs that brush as these will cause 'wear and tear', and you can assess these faults from the ground.

Now see the horse being ridden and again look for his general attitude to work as well as the way he goes. See him perform at all paces and over some jumps. Look at the rider to see if he is skilfully masking some faults, and note how he has to ride the horse generally; is he having to work hard to get results, and what tack and artificial aids is he using? Under saddle, take another mental picture of how the horse looks. His quarters should not appear higher than his withers, or he will have difficulty in getting his hocks underneath him. Look for the activity in the hocks and shoulder; are they supple and used freely?

The movement of the horse's tail can tell you a great deal about the way he is using his back and about his general attitude. A tail which is held away from the quarters and appears relaxed and swinging indicates that the horse is using his back correctly. But don't confuse a swinging tail with one that swishes, for this indicates resistance. The tail held stiffly or clamped in may indicate a problem with the horse's back which will interfere with his movement.

See the horse ridden out of the yard and away down the road, then back and past the yard entrance. You don't want a horse which 'naps' and won't leave home. Try to see how he reacts to traffic—you don't want a traffic-shy horse. Then it's your turn to ride and you will really be able to assess how the horse feels and whether you suit one another. Perhaps you can ask him one or two extra questions to test his response; how does he respond to *you*?

The essential vet

If the horse passes your test and you make an offer, make it clear that this is subject to a thorough veterinary examination. Have the animal examined by your own vet or an independent vet of repute: obviously it is not the thing to ask the seller's vet to examine it. And if you're having to employ a vet outside your area, make sure he is a specialist horse vet. A thorough veterinary examination will take a good two hours by the time respiration and pulse have been tested after strenuous work, and it is of course a service for which you will have to pay, but in view of the money it may save you it is well worth the insurance. Speaking of insurance, *do* insure your horse for a realistic sum, for which you will need a vet's certificate anyway. In any event, you ought to have a third party liability cover and in Britain this is automatic for members of the British Horse Society.

Incidentally, when a vet passes a horse 'sound in wind and limb', he will qualify his statement with

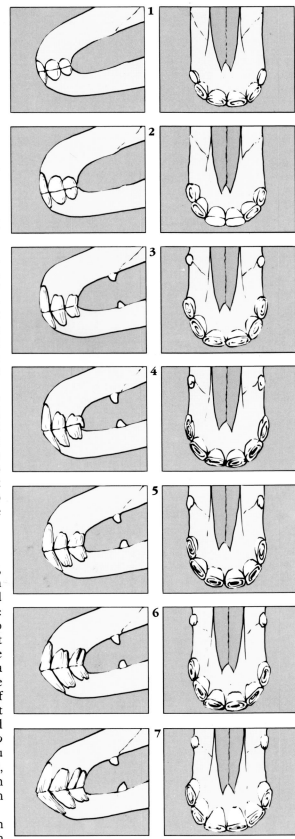

Left By looking at the shape, wear and colour of a horse's teeth a prospective owner can get some idea of its age.

The six pairs of incisors (front teeth) of a yearling (1), which are temporary, or milk, teeth, are distinctively white and are straight.

By the age of three (2) the central incisors are permanent and show a little wear on the biting surfaces. The incisors of a six-year-old (3) are all permanent and show definite signs of wear with dark cavities or 'cups' on the biting surfaces. The corner incisors touch evenly, and the male has developed the tushes, or canines.

The corner teeth of a seven-year-old (4) have developed an obvious hook and the cups are less distinct.

By eight years of age (5) the hooks have worn away and a dark line has appeared under the cups of the central teeth.

A ten-year-old (6) has developed a dark hollow, known as Galvayne's groove, on the corner teeth and the central teeth have become triangular. The cups on the biting surfaces are starting to disappear and the teeth have a definite slope.

The teeth of a 20-year-old (7) noticeably protrude. The Galvayne's groove now appears down the whole of the corner teeth and the biting surfaces have worn completely smooth and have lost the cavities.

the purpose for which the horse is intended. So do make sure that he knows why you want the horse and he will assess him accordingly. For instance, a horse which is perfectly 'sound' for gentle hacking is not necessarily suitable for hunting.

If you are going to keep your horse at grass, you should consider having him freeze-marked or branded for identification in case he is stolen. In all cases, make sure you obtain all the necessary papers or breed certificates from your seller— particularly if you are buying a continental horse— and have the details changed with the relevant societies. Ask for vaccination certificates and details of regular medications like worming.

An owner's obligation

From the moment you buy your horse you will have certain legal rights and obligations, as well as a moral commitment, and it is as well to be aware of these.

The subject of horses and the law is complex taken in its entirety, and there will be differences, not only from country to country but also from state to state, territory to territory, county to county. Therefore, while these notes will give an idea of the areas of horse ownership which involve the law, if you are concerned specifically with any one you are recommended to take local legal advice. Your National Federation should be able to help in many instances as it is their responsibility to defend the rights of horse owners and riders in the courts, where cases involving rights of way, planning authorities, traffic accidents, etc., are concerned.

Caveat emptor

For the private owner, the greatest concern with legalities probably comes with the purchase of a horse. The veterinary certificate as already de-scribed is a good assurance against the ancient law of *caveat emptor*—let the buyer beware—which, unless he takes such precautions, could leave a gullible purchaser saddled with a real problem. The onus is on the buyer to find out all he can about the horse: the seller need not divulge anything. On the other hand, if he does specify, even verbally, that the horse is suitable for a certain purpose, for which it subsequently proves totally unsuitable, then the buyer may have just cause for redress. These legal aspects become very technical, but it is as well to be aware that if you do suspect that your purchase is not quite as described in the advertisement, or as the vendor led you to believe, it might be a good idea to seek legal advice. Unfortunately, this could turn out to be rather expensive. In some instances the vendor will be obliged to take the horse back, or at least pay costs to the purchaser. You should be careful of a seller's 'conditions' or 'warranties', particularly in the sale ring, where a horse may also be for sale with an existing veterinary certificate: no matter what it

The walk *This is a pace of four-time, with four distinct foot falls heard to each stride.*

The sequence of hoof beats shown above is off-hind, off-fore, near-hind and near-fore.

The trot *With a pace of two-time, the horse advances on diagonals – near-hind and*

off-fore, and off-hind and near-fore. At the rising trot the rider should sit on the left

The canter *This is a pace of three-time. The horse, seen above with the off-fore*

leading, strikes the ground with the near-hind, the diagonal off-hind and the near-

The gallop *A fully extended pace of four-time, when, as in the canter, the horse*

'leads' with one or other foreleg: the sequence for off-fore leading, being near-

The horse should move with long even strides in regular rhythm, with the hind feet

falling in, or in front of, the imprints left by the forefeet.

diagonal on the right rein, as the near-fore contacts the ground, and the right

diagonal on the left rein, when the off-fore comes to the ground.

fore together, followed by the 'leading leg' to each stride. If four foot falls are heard

this could mean a lack of balance or an attempt to collect a young horse too soon.

hind, off-hind, near-fore, off-fore and suspension. The moment of suspension

(above) shows how wrong old sporting pictures were in depicting horses at full stretch.

costs, you should commission your own veterinary examination.

Third party liability

Once you have your horse, the greatest liability is that to third parties, and as already mentioned your first obligation is to make sure you are properly insured against accidents or damage involving third parties. Nowadays, road accidents involving horses are horrific both in quantity and description, often as the result of the modern motorist's ignorance of the horse. Although riders have a legal right to use the highway, they also have a moral obligation to use it sensibly and courteously, thanking careful drivers. Make sure you are well protected and easily visible, especially in poor light, by wearing a reflective belt or vest, for example, and observe the rules of the road.

It is also, of course, an owner's duty to ensure that gates and fencing are secure. Again, you may be liable for any damage your horse causes by breaking loose onto crops or the road. It is worth bearing in mind also that if you borrow a horse, you may be liable for damage caused by it to third parties, and you will have no redress against the lender if you are injured in using the horse for a purpose for which it was not lent, but he should have let you know of any vices or defects the horse may display for the agreed purpose. If you do borrow a horse for any length of time it is as well to have a proper legal agreement drawn up.

If you are proposing to keep your horse at home, and build your own stables, you should of course check on local planning restrictions. You could waste a great deal of time and money in building stables illegally, subsequently to find you must remove them. It is as well to be sure that you will not receive complaints from neighbours which could lead to restrictions or even a prohibition to you having your own 'yard'.

If you are keeping your horse at livery, you may have to enter into a contract with the yard owner and you should of course be clear on every point before signing an agreement. Most yards will display a disclaimer of liability, but it is as well to remember that if damage or accident can be proved to have been caused by their negligence then the disclaimer is not worth the paper it is printed on. If your horse has any particular vices you should of course make the yard owner aware of them so that he knows what he is taking on and can instruct his staff accordingly.

A moral obligation

Your moral obligation to your horse is to keep him for the benefit of his own welfare. The next chapters in this section give you guidelines to follow and build up your standards of, and expertise in, horse management, but practical experience is the best tutor. Your horse's condition will reflect your standards.

2

STABLE
MANAGEMENT

Without a sound, healthy body and mind a horse cannot work to best advantage. An unhealthy, unfit horse put to work is a danger to himself, his rider or driver and others with whom he comes in contact. The demands we make of a domesticated working horse far exceed those with which he would be faced in the wild; therefore, to enable him to carry out those demands effectively and safely, to the best of his ability and training, a high standard of care and management is essential.

The basic principles of stable management follow the lines along which the horse evolved in the wild. He is, by nature, a herd animal, and all but a very small minority of horses are happy when in the company of other horses and unhappy and insecure when alone. So, although a horse must learn to work alone and be alone at times, as a general rule he should be kept with others. He needs adequate exercise and, as wild horses graze for eighteen to twenty hours a day, should be fed little and often, so that a small amount of food is passing along his digestive tract all the time, and have a reasonably bulky diet to resemble his natural diet of grass. Add to these basic needs a final one of shelter, and you have in a nutshell the four main areas of horsemastership which need to be attended to if you are to keep your horse fit and happy.

The basic facilities required to keep a horse are a stable, storage facilities for feed, bedding and tack, and a paddock or field for freedom and grazing.

The stable Horses are big, strong animals, and need fairly big, strong buildings to house them. A horse of, say, 15hh will need a stable of *at least* 10 × 12 feet (300 × 365 cm) to enable him to turn and walk round in comfort and stretch flat out to sleep; a standard full-size box is usually 12 feet (365 cm) square, and the bigger the better. If the box is too small, the horse will be cramped and, when lying down or rolling, could become cast (trapped up against the wall), in such a position that he cannot rise without human assistance. Horses panic easily and, in the ensuing struggle, can seriously injure themselves trying to escape from a

cast position; a stable of adequate size is one method of helping to avoid this, as is an adequate banking of bedding around the stable walls.

Ventilation is most important to a horse's respiratory health, so the stable must allow constant circulation of fresh air, normally by means of window and door, and, if possible, by means of ridge roof ventilators and louvre fitments in gable ends. A healthy flow of fresh air is *not* the same as a draught, which should be guarded against at all costs. A conventional stable in a modern, and sometimes not so modern, yard consists of a loose box, the door of which has a lower and upper half. The upper half should be kept open in all but the worst weather, such as driving snow or rain, or when the wind is blowing directly in, which will ensure not only some exchange of air but also that the horse can put his head out and so breathe fresh air and see what is going on in the yard. In cold weather it is better to add extra blankets rather than shut up the door, unless absolutely necessary.

It is usually recommended that stable windows open into the stable in an upward direction to direct the flow of air over the horse's back and so not create a draught, but it is obvious that if the top door is open any niceties concerning the design of the window are largely obviated! However, there are times when the door must be shut and windows should be guarded with strong metal grilles which must follow the line of the window to allow it to open, but guard against presenting a projection against which the horse can knock his head.

The height of the stable is also important, not only from the point of view of ventilation but also of safety. A rearing horse can reach a height of up to 12 feet (365 cm), depending on his size, and would be in danger of banging his head on a roof or ceiling lower than this. Although most horses live happily for years in much lower stables than this, it is an ideal to be aimed for. As a minimum, for safety reasons, the eaves height of a stable should be 7 feet 6 inches (230 cm). The door should be a similar height and at least 4 feet (120 cm) wide and should be securely bolted.

Electric light fittings should be well out of reach of the horse's enquiring teeth or strongly guarded, with wiring inside metal conduits and waterproof switches outside the stable. Stable fittings must not cause a possible hazard to the occupant, so use triangular mangers which fit into a corner, and automatic waterers or buckets (two per horse) sited in corners; the former, although efficient most of the time, must be kept clean and they give no indication of how much the horse is drinking—an important guide to his well-being. The stable should have at least one strong ring or bracket (bolted through the wall) at head height to which the horse can be tied for grooming or while mucking out. His haynet can also be tied to it. If you prefer a hayrack for hay, use the corner type again, with the top set at slightly above head height, so the horse can reach it comfortably without hitting his head or having bits fall in his eyes.

The construction of the stable must be strong, with good insulating properties and the building should be sited with its back to the prevailing wind. Stone and brick are the best; solid (or almost solid)

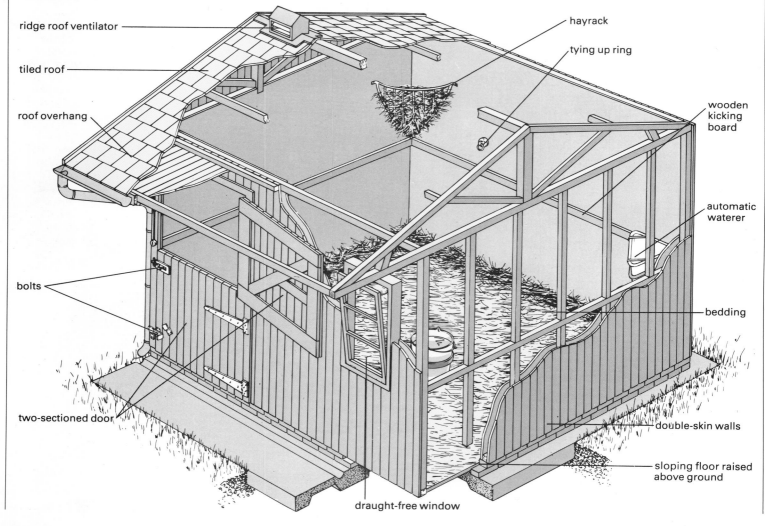

ridge roof ventilator

tiled roof

roof overhang

bolts

two-sectioned door

draught-free window

hayrack

tying up ring

wooden kicking board

automatic waterer

bedding

double-skin walls

sloping floor raised above ground

concrete blocks (not compressed ash breeze blocks) can be good if lined with wood to allay dampness, and timber is very common and good if double-skin walls are used. Any stable, even brick or stone, is improved by wooden lining boards (planks or 5-ply panels), at least round the lower half, for protection of the structure against a kicker, and extra insulation.

The roof should be of material that is a poor conductor of heat. Metal is therefore not advisable, but if its use is unavoidable it must be lined with wood or industrial polystyrene panels. Slates or tiles are excellent, and double-skin wood under roofing felt is good. A roof overhang of 3–4 feet (90–120 cm) shelters the stable from wind and rain, and provides extra shade in the summer.

Few stables today have the special stable-brick floors of the past, grooved for drainage and very hard, which are superb if you can afford them. Asphalt is quite good in cooler climates as it is warm and not too hard on the legs; it is affected by heat, however, and is slippery when wet. Concrete, although very common today, is not particularly good. It is extremely hard and cold, even through bedding, and it holds moisture and can be slippery unless the surface is roughened. When used, bedding should be down for warmth and protection whenever the horse is in. Special interlocking, perforated plastic tiles can be obtained and, laid over concrete, make a good-draining floor. They contain 15 per cent rubber to prevent slipping, and can be laid direct onto a deep foundation of gravel and sand. Concrete—not breeze—blocks can also be set into a similar foundation with drainage channels of an inch or two left between the blocks, and this makes an efficient floor. Old-fashioned cobbles are hopeless unless kept well covered and are very difficult to clean. All floors should be laid on good foundations and raised slightly above ground level.

Although in theory drains can be dispensed with, unless necessitated by by-laws, the more efficient the drainage the more economical will be the use of bedding, so much of which is often discarded purely because it is wet. If building new, make sure the stable is on firm, dry foundations, slope the floor slightly from front to rear towards an outlet point, and make sure the wet does not pool inside the box, or you will find your bedding bills very expensive.

Storage areas You will need somewhere secure, dry and reasonably warm to keep your tack. Leather rots in a cold, damp atmosphere, as do rugs. The tack room is also used to store tack cleaning equipment, grooming kit, veterinary supplies and many other general items, so should be fairly central to your operations, and preferably have hot and cold water laid on for tack cleaning.

Feed and bedding need airy, dry storage areas if they are to keep well. It is hard to keep vermin out

of straw and hay, but other feeds should be kept in galvanized metal bins with lids tied firmly on. Plastic dustbins can be used, though rats can gnaw through plastic. Never keep your feed in its sacks if you want to protect it from vermin and damp. Feed rooms should have facilities for boiling linseed, barley and water, although for one-horse owners this can be carried out in the kitchen at home.

The field Although many horses—notably army, police and town delivery animals—live contented, healthy lives without ever being turned out, the chance to graze and play in a paddock, particularly in company, is a mental and physical tonic to a horse, and every effort should be made to provide this facility.

For a stable-fed athletic horse, the quality of grass is less important than the safety of the field. It must not be excessively hilly or rough, and must have no objects on which the horse could injure himself, such as farm machinery, discarded wheelbarrows, harrows hidden in the grass or general litter; something to pay particular attention to if a foot-path goes through or alongside the field. Ponds, unless shallow and clean, are usually safer fenced off, and the fencing itself should be smooth and strong. If there is no thick, natural hedging round the field, timber posts and rails are the next best thing, although expensive, and should be constructed to at least 4 feet (120 cm) high. Smooth wire strained taut between wooden or concrete posts is good, and there are some proprietary designs of fencing comprising flexible plastic or rubber strips between posts, which are also good.

Barbed wire is the very worst fencing for horses, who can easily run into it and sustain ruinous injuries. If you are renting accommodation for your

Above An ordered modern yard. The wide overhang protects the occupants from bad weather and offers shade. The doors and surrounding areas are used with metal to prevent horses from chewing wood.

Above left Horses stable quite happily in stalls, provided they have enough halter rope to allow them to turn their heads and lie down.

Below left A typical timber-built loose box. This should measure at least 12 ft (365 cm) square and 7 ft 6 in (230 cm) high, with a minimum door width of 4 ft (120 cm). A window, and if possible a ridge vent or louvre fitment in the roof and gables, is essential for ventilation. A hayrack and automatic waterer are optional; hay can be fed from a net and water from a bucket.

Overleaf (page 21) Always protect a horse well against injuries when travelling.

horse and have no choice of fencing, consider investing in your own set of electric fencing to erect inside the existing perimeter, then you can take it with you when you move. Demonstrate it to the horse by wetting his nose and pushing it against the wire, when he will get a shock just unpleasant enough to keep him away in future!

Gates, too, must be safe and strong—wooden barred or tubular metal—and kept padlocked at *both* ends for security.

Water troughs can constitute a hazard if they have sharp edges or corners, but can be rendered safer by tying straw bales or old car tyres round them. Taps should not protrude above the top of the trough. If there is no water laid on, supply it in plastic dustbins tied to the fence top and bottom and filled by hosepipe. Buckets, however, would do if the horses are only out for an hour or two.

Management systems

From the horse's point of view, there are three methods of keeping him:

1 Stabled.

2 At grass.

3 On the combined system, whereby he spends part of his twenty-four hours stabled and part at grass. Here the horse gets the best of both worlds, enjoying the shelter and comfort of a stable plus freedom, grass and, hopefully, the company of other horses out in the field.

The stabled horse's diet and life is completely controlled by man. He can be made supremely fit by controlled exercise, feeding and grooming. He is always handy when needed, kept clean and trimmed, is often clipped and rugged up in winter—and takes a lot of time to look after.

The combined system, however, has much to recommend it. Horses can be kept fit on it and the grass helps keep their digestion in order if they are on a heavy diet of concentrate feeds. The time in the field also helps relieve boredom from standing in. In winter, the horse can be stabled at night but out for all or part of the day while his owner is at work. In summer, he can shelter from the heat and flies in his stable during the day and be turned out in the cool of the evening. If the grass is to supplement diet, a field of 2–3 acres (1–1.5 ha) is recommended, but size is not so important if the horse is out just for exercise. Droppings must be regularly removed, however, and long grass cut. Another advantage of the system is that the horse can exercise himself to some extent, so lessening demands on a busy owner's time.

From the owner's point of view, there are also three ways to keep a horse, apart from employing staff to do so:

1 By looking after him yourself, either at home or in rented accommodation.

2 By keeping him at full livery (boarding him out) and paying the stable owner and his staff to cater for all the horse's needs (excluding farrier and veterinary care).

3 By keeping him at part livery, paying for stabling and grazing and some agreed services, and doing the rest yourself; or working livery when the stable owner has the use of your horse for a number of hours each week.

System 2 is obviously the most expensive and least time-consuming, with system 1 at the other end of the scale and system 3 in between.

Only you can decide how much time and money you can allocate to the horse to ensure that he is adequately cared for and happy.

Feeding and watering

Feeding (including watering) is the single most important topic of horsemastership. Your horse *is* what he eats if he has an adequate amount of good-quality food which contains the right balance of nutrients and is fed in a sensible, rationed way, you are at least half-way to a healthy, contented horse fit to work on. A horse who is underfed or given poor-quality food of low feed value will never be well or achieve his full potential, while an overfed one is in

constant danger of serious health problems, sometimes with lasting effects, and can become dangerous to handle and ride.

Purposes of food Food is used by the body for:

1 Maintenance of body temperature—about 100.5°F (38°C) in the horse.

2 Formation and replacement of body tissue (skin, muscle, etc.).

3 Providing energy for life processes and movement.

4 Putting on weight (storing excess food as reserves of fat).

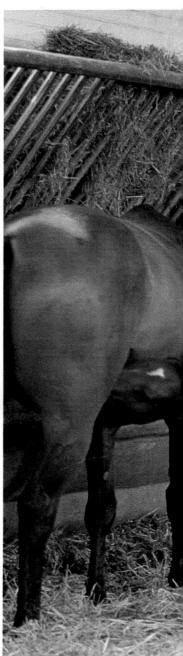

Below left Car tyres or bales of straw tied round a water trough prevent a horse injuring itself on any rough edges.

The first two are the most vital to life, and if there is insufficient food and water to maintain them the horse will die. If we work a horse and/or expose him to cold weather, his body will demand more food to cope. If the food (or fuel, if you like) does not arrive, the body will use its reserves and the horse will lose weight. If the reserves become depleted but the demands are still made, the body will draw on its own tissues for fuel. Muscles will waste away; the horse will become emaciated. Ultimately, the body temperature will drop below a workable level and the horse will die.

Conversely, if too much food is eaten, the horse first puts on weight. He becomes obese and a prime candidate for crippling disorders like laminitis in the hooves, azoturia, strained ligaments and tendons, lymphangitis and colic.

In the final analysis, despite the scientific formulae on feeding which abound, it is still the knowledge, skill and judgement of the person who looks after him which will determine a horse's bodily condition.

Water The horse's body is about 70 per cent water. Very roughly, a horse could need between 6 and 12 gallons (27 and 54 litres) a day, depending on work and weather. Water is needed for the

Below Mares and foals, on a German stud, housed loose.

production of all body fluids (such as blood, lymph, digestive juices and milk) and for the excretion of waste products in sweat, urine and faeces. Without enough water the body becomes dehydrated—a horse may die within a few days without water.

Types of food There are two main types of food—fibrous foods such as hay, and hay-age (grass conserved to a stage midway between hay and silage, useful to replace hay for horses allergic to it), and concentrate feeds like oats, barley, flaked maize and cubes or nuts. A food containing less than 20 per cent fibre is classed as a concentrate. Bran, although strictly a concentrate, is not used as such as it is rich in phosphorus and low in calcium, causing a mineral imbalance resulting in bone disorders. It is useful for making easily digested mashes for sick or tired horses and as a base for mixing other foodstuffs, and should always be dampened. Roots such as carrots, turnips, apples and sugar beet pulp add succulence and variety to feeds as well as providing the important vitamins present in fresh foods.

Fibrous foods of good quality (hay and hay-age) with a protein content of 11 or 12 per cent are adequate alone for a stabled horse doing light or no work in mild weather. Harder work and cold weather demand a 'higher octane' fuel, a more concentrated, energy-giving food, such as oats, barley or cubes, in addition to the fibre. Generally, the harder and faster a horse works, the more concentrates and the less fibre he needs, although fibre should never fall below 50 per cent of the total ration. For a horse doing reasonably active riding club type work, about two-thirds of the daily weight ration should be hay.

How much? A very simple but effective way of determining roughly how much food a horse is likely to need each day is to double his height and express the answer in pounds. This will be slightly on the generous side, so subtract a few pounds, depending on whether the horse is a good doer (makes good use of his food) or a poor doer (difficult to keep weight on). So, for a 16hh horse, twice 16 is 32 = 32 lb, say 30 lb, of which 20 lb could be hay and 10 lb concentrates. (When working in kilograms, expressing the horse's height as 10 cm = 1 kg will give roughly the same result.) This works only for horses of 15 hh and above. Ponies and hardier cob-types need less concentrate food. Count bran as a concentrate, but because of the previously mentioned imbalance it can cause, give no more than a quarter, and preferably a sixth, of the concentrate portion as bran.

Rules of feeding: 1 The best-known rule is to feed little and often. If a horse is due to work hard and fast, his hay should be taken away two hours beforehand; one hour for light work. After work, he can (and probably should) have hay with him all the time to chew on at will. Concentrates take longer to digest than hay, and

3–4 lb (1.5–2 kg) in one feed is enough; therefore, if he is on 10 lb (5 kg) of concentrates daily, split it into three feeds, giving the largest at night so he has time to digest it.

2 Water before feeding to lessen the chance of large amounts of water washing particles of undigested food from the stomach into the intestines. If there is always water there, however, horses drink moderately and can safely be left to their own devices.

Left Lack of condition should be apparent before a horse reaches this sorry state—correct feeding is paramount.

a   b   c

Left Vitamins and minerals (a), tallow (b)— a high-energy fat—, and molasses (c)—a ready form of energy—are used in horse and pony cubes to create a perfectly balanced feed. Vitamin supplements can be added to other concentrated feeds to ensure a balanced diet and molasses, a syrup made from sugar, make feeds more palatable.

Below A range of concentrated feeds. It is best to buy whole grains and then freshly prepare the feed by crushing the grain or cooking it. Wheat and oats can be bought as a meal and beans and seeds, which should be fed in moderation, are available as a meal or in flakes or pellets. Cubes and nuts, which provide the ideal balance of nutrients, are manufactured for the specific requirements of the type of horse or pony.

From right to left:
Grass nuts
Wheatfeed meal
Locust-bean meal
Whole oats
Soya bean meal
Whole maize
Sunflower seed
Oatfeed meal
Whole barley
Linseed cake

3 Make all changes in type and amount of feed gradual to give the digestive system a chance to adjust, otherwise indigestion (colic) could result.

4 Feed only good-quality food. Bad food is uneconomical as it has low nutrient levels, and horses often leave it anyway. It also poses a health hazard. Any food which smells sour, musty or generally unpleasant, which is dull, dusty and dirty or, worse, has white mould on it, must be avoided. Hay must smell sweet and spring apart when you cut the binder twine. Grains must taste pleasant and wholesome and look clean and bright, and bran must be light, flaky, sweet, and not lumpy.

5 Adjust food according to temperament and body condition, taking into account the work being done and the weather. If the horse is too excitable or too fat, cut back the concentrates and increase the hay and, if possible, the work. If he is losing weight and/or is dull and listless (but healthy) feed more concentrates.

6 Do not work fast or hard for at least an hour after a full feed. The stomach lies next to the lungs and when it is full of food (including grass) the functioning of both will be hampered.

7 Keep to the same feed times each day to give the digestive system a routine, and do not leave gaps of many hours between feeds. Leaving hay always with the horse will lessen the effects of this if it is ever unavoidable, or the horse only needs one or two feeds a day.

Feed supplements Vitamin and mineral supplements can be invaluable in making sure your horse has a correctly proportioned supply of all essential vitamins, minerals and trace elements. Your veterinary surgeon or an equine nutritionist will help you choose a suitable supplement for your horse's needs. Salt should always be available in the form of a lick block in the stable.

Preparing food Grains are best fed bruised or rolled, which just cracks them open to allow entry to the digestive juices, or cooked and flaked. Oats are usually fed bruised; barley is fed bruised, boiled or cooked and flaked, and maize is cooked and flaked. Ideally grain should be bought whole and rolled freshly for feeding: the feed value of rolled grain deteriorates after about three weeks.

Oats and barley are staple foods, both high in protein and carbohydrate. Barley causes horses to be less excitable than oats and can be interchangeable, although its higher level of carbohydrate makes it unsuitable for faster work. Cubes or nuts are a good choice for the amateur owner as they offer a balanced, compound feed which can be interchanged in whole or part for the grain ration. They are low in moisture, however, and adequate water must be available. Maize, being almost entirely carbohydrate, is not sufficiently balanced and should not comprise more than a quarter of the total concentrate ration, although it is most useful for thin animals. Beans can be fed to horses at grass in winter, but they are very high in protein, and

Studs can be fitted into shoes to give a competition horse more confidence if the ground is hard or slippery. The

studs screw into holes made in the shoes by the farrier and are then easily removed after the competition and the

holes re-plugged. This sequence of four photographs show how studs are fitted into the horse's shoe. From left

to right, the shoe is tapped to accept the correct stud which is screwed in and tightened using a special spanner.

thus very heating and fattening.

Feeds should be weighed accurately on a scale, put in a bucket and *thoroughly* mixed. A little bran aids distribution, especially where a powdered feed supplement is used, or with cubes to promote mastication and salivation, and should be dampened either with plain water or perhaps with diluted black treacle or molasses until the feed is just crumbly. The horse can eat either from the bucket or from his manger.

Cooked feeds and mashes are ideal for sick or tired horses. To make a basic bran mash, fill a feed bucket half to two-thirds with bran and pour in boiling water, mixing thoroughly until the bran is crumble-damp throughout. Cover and wrap the bucket with an old blanket and leave it to cook and cool, then mix well by hand to check it is cool enough to feed. Basic mashes are unappetizing to eat, but can be made interesting by adding black treacle, coarsely grated carrots and a handful of salt.

Cooked grain can be added to a mash or other feed. To cook barley or oats, use uncrushed grain, soak in cold water overnight and boil in water (or steam in a sieve) till soft.

Boiled linseed can be used to make a mash instead of water. Take a teacupful of the grain, soak in cold water for twenty-four hours, then boil hard for at least twenty minutes to kill off the poison which can occur in improperly cooked linseed. Then mix with bran as usual and allow to cool. Linseed is very high in fat and is good for horses in

Above left *The correct steps to tie up a haynet. It should never be tied so low as to risk the horse's legs getting entangled in it.*

Left *Salt should always be made available in the form of a lick block.*

Right *If you are competing, take a feed for your horse when it is relaxing after the event, and provide hay for the journey home.*

poor condition; it is also fed occasionally to fit horses as a laxative.

Roots should be sliced up thinly (carrots lengthwise to avoid forming chunks which could choke the horse), and sugar beet pulp, which is good for keeping weight on in winter, must be soaked in two and a half times its own volume of water for twenty-four hours before feeding, otherwise it will swell inside the horse and could cause choking or rupture his stomach or intestine. Once soaked, squeeze out a large double handful or two and mix with the feed thoroughly. Sugar beet pulp is not suitable for horses doing hard, fast work. Ideally a stabled horse should have up to 4 lb (2 kg) roots daily. Horses on grass do not really need them.

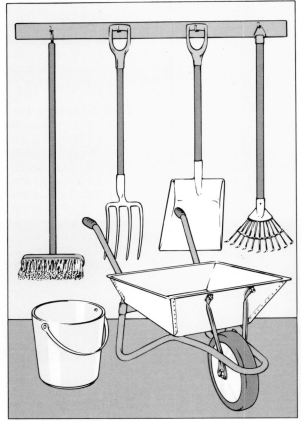

Bedding
Bedding is used to help keep a horse warm and clean, to encourage him to lie down and rest properly, and to protect him from the hard floor when he does so. A thick, dry, clean bed adds enormously to a horse's comfort and well-being, to the extent that it can lessen the amount of food needed.

Materials The best and most common materials are straw (usually wheat but sometimes barley or oat), wood shavings and shredded or diced paper. A good-quality dry peat or tan bark is a good

bedding if carefully managed and is of course very easy to dispose of after use—an important consideration. Guard against dusty peat, however, and for the same reason sawdust on its own is not satisfactory.

Straw, when fluffy, clean and golden, looks nice and is still extremely common. Even top-quality samples, however, can contain spores and fungi invisible to the eye but which cause an allergy known as broken wind (emphysema or chronic obstructive pulmonary disease) in susceptible individuals. This condition produces symptoms akin to asthma in humans, and effectively writes off an equine sufferer from an athletic life. Some horses also eat straw either from boredom or because they are feeling queasy from too many concentrates (like humans eating a heavy, rich meal), and nowadays one has to bear in mind that residue of insecticides and weedkillers may remain on the straw and could be harmful.

Shavings are inedible and less dusty than straw can be (especially vacuum-cleaned shavings), but even they can cause allergic skin responses in some horses from the chemicals with which trees are now often sprayed. It is not quite as easy to keep beds clean as with straw, as the droppings break up

Below left Tools for mucking out. Dirty straw bedding requires a manure fork and shovel, a barrow, a bucket or skip, and a broom to sweep the floor. Shavings are managed well with a wire rake, while soiled paper bedding can be removed by hand, wearing rubber gloves.

Below Boxed up and ready to go. Check all necessary equipment is aboard. Most horses travel best facing away from the direction of movement. Legs need protective bandages, with gamgee padding covering the heel and coronet, and also knee and hock boots, if tolerated. The tail should be bandaged and a poll guard protects the head. If chilly, a day rug completes the picture.

and mix in easily with the bedding, but with care, shavings, possibly mixed with sawdust, can make acceptable bedding.

Shredded or diced paper is increasingly popular and ideal for horses prone to allergies as it is non-dusty and chemical free. It is easy and light to work with, dries out quickly and is inedible. It is not expensive if you resist the temptation to remove more than the really dirty, sodden stuff.

Systems There are three systems of managing a bed:

1 Full mucking out (suitable for straw).
2 Deep litter (straw, shavings).
3 Semi-deep litter (any material).

In mucking out, the droppings and very dirty, wet straw are taken out, the clean and part-clean straw separated, the floor swept and the bedding

replaced, with fresh bedding added. This is a fairly time-consuming method.

In deep litter, the droppings only are removed, and fresh bedding is laid on top of the existing material. You need a well-ventilated, dry stable for this system, which works better in winter than summer as it can attract flies in some situations. Provided droppings are meticulously removed, though, and the old bedding left undisturbed to metamorphose into compost, it makes a warm, springy bed for the horse and will be dry, crumbly and without smell when removed several months later. This is a quick method on a daily basis.

Semi-deep litter, a cross between the two, is suitable for stables not ideal for full deep litter; the droppings are removed together with the most badly soiled bedding. The resulting space is filled in with semi-clean material from the banks around

the sides of the box, and fresh bedding is put on top. With shredded paper, the bed is tossed and turned to dry out before replenishing with new material.

With any system and material, the secret of successful bedding is to remove droppings as scrupulously and as often as you possibly can.

Grooming

A dirty horse is neither prestigious nor pleasant to work with, nor will his clogged skin be working to best advantage to help regulate his body temperature and help eliminate metabolic waste via the sweat.

Stabled horses need grooming to promote the improved efficiency demanded of the skin by an unnatural environment and higher than natural levels of work and food.

Grooming is best carried out after exercise when the skin is warmed up and the pores opened, enabling the easier removal of excess grease and dandruff. On return from exercise, mud and sweat should be washed off immediately with cold water and this will make grooming much easier. Make sure heels are dried out properly. Many horses enjoy a cold hosing over the whole body in hot weather or if very hot and sweaty.

Stable stains should have been removed with the dandy brush or cactus cloth before exercise, but the same brush is used to remove dried mud before you get to work with the body brush, although a rubber or plastic curry can also be used.

Begin body brushing at the head and brush in the direction of the hair. Work from front to back and top to bottom so you do not brush dirt onto already clean parts. Lean your weight onto the body brush, with your arm stiff but slightly bent, and use long, sweeping strokes, getting through to the skin. Every three strokes or so, run the bristles firmly over the metal curry comb to clean them, and tap the dirt out of the comb onto the ground outside the box. Go gently round bony or sensitive areas, and pay particular attention to the areas under the forelock and mane, between the legs, under the belly, under the tail and behind the pasterns—the areas which most harbour dirt and are most often overlooked. To promote muscle tone use a wisp with a banging action on the flat areas of neck, shoulder and quarters—a form of massage. Build this up gradually though, to avoid making the horse sore.

Brush out the mane and tail (not forgetting the forelock) with a body brush, and, if necessary, use a metal comb to very gently pull, or thin out, the mane and tail hair.

Next, damp-sponge the eyes, nostrils and lips with one sponge and the sheath and under-tail area (dock) with another—and always keep the sponges separate.

Pick out the hooves with the hoofpick, working

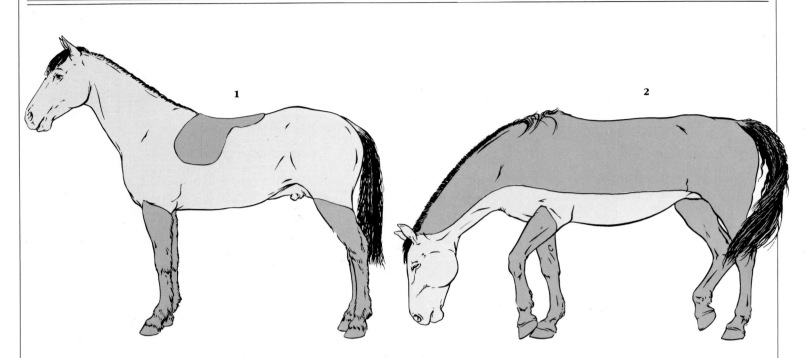

from heel to toe, not forgetting the sides and centre (cleft) of the frog, and oil the walls and sole with hoof oil.

Give a last dusting over with a damp stable rubber, and you have finished.

Shampooing If your horse is very dirty or you are attending a special event, you might want to bath him. Use warm water and an animal shampoo and douse him gently (avoiding the head) with a large sponge. Work quickly and thoroughly. Rinse by hosing down, really getting out all the soap, and rub dry as quickly as possible with straw or old towels. Walk the horse round to dry off completely. Make certain his heels, pasterns, ears and loins are really dry before rugging up normally, according to the time of year. Avoid bathing altogether in chilly or cold weather.

Clipping A working horse in winter will need clipping unless his coat is exceptionally short and fine. Clipping in the latter part of the winter should be avoided for fear of damaging the emerging summer coat. If the horse is to be out during the day, a trace clip will probably be sufficient to prevent him from becoming over-hot and wet in work while offering reasonable protection in the paddock, although a New Zealand rug will be needed on cold days.

The illustrations show the various standard clips. Basically, the longer and thicker a horse's coat and the harder he works, the more hair should be removed if he is not to become chilled and lose condition through excessive sweating. But remember, steeplechasers can be made racing fit with only a trace clip or chaser clip. If these clips are good enough for them, they are good enough for less hard-working animals.

3   4

Basic tack and clothing

Above The standard types of winter clip. The hunter clip (1), used for fast, hard-working horses, leaves only the saddle patch to prevent saddle sores and the leg hair for protection, especially while hunting. The chaser clip (2) suits fast horses which tend to get cold easily and whose coats are not too thick. The blanket clip (3) removes all but the leg and back hair, which in mild weather can act as a substitute rug. The trace clip (4) is ideal for a horse kept out during winter and the hair may be left on under the throat.

Left A horse appreciates a refreshing wash down after hard exercise but make sure it is then covered with a sweat sheet and walked around to help it to dry.

Right To tie up a horse safely at a competition, link the halter rope to a piece of string attached to the trailer.

Rugs and blankets The clipped horse in winter will need clothing when not working to replace his coat and keep him warm; probably at least a top rug and one under-blanket, maybe more. To check if your horse is cold, feel the base of his ears—if they are cold, so is he.

Today, when time and money are at a premium, synthetic clothing (quilted nylon rugs and Acrilan blankets) which can be laundered in a day while the horse makes do for a few hours with old reserve clothing, enables owners to get by with one good rug and one or two blankets, although two sets make life easier.

In summer, a summer sheet (linen or cotton rug) is a dispensable item which does, however, keep the coat clean and smooth, and keeps flies off the body.

Headcollar A headcollar, or halter, made of leather or nylon webbing is needed to lead or tie up the horse, together with a leadrope.

Bridle A simple bridle with a snaffle bit is adequate for most horses, at least as a starting-off point. The bit should protrude just ¼ inch (6 mm) from the mouth at each side for correct width and comfort. The bridle itself must not cut into the base of the ears or rub the cheekbones, and you should be able to fit your fist between the throatlatch and the horse's jowl.

Saddle After your horse, your saddle will be your most expensive purchase and most people can only afford one. A general-purpose saddle is most people's choice for varied riding, as it is suitable for jumping, hacking, and even showing and dressage, as it has moderately forward flaps. A spring tree adds greatly to the rider's comfort and correct fit to the horse's back. A good saddler will fit a selection for you, but the points to watch for are that the saddle must have four fingers' width of clearance from the withers to the front arch with the rider in the saddle, and a similar distance at the back, with a clear tunnel of daylight along the horse's spine, upon which there must be *no* pressure. You should

just be able to slide the flat of your fingers in the sides of the front arch to check the width. When the horse moves, check that the flaps do not hamper shoulder movement. A numnah, or saddle pad, should always be worn with a spring-tree saddle as the natural movement in the saddle could cause sores.

Bandages Woollen stable bandages are useful for drying off legs and are equivalent in warmth to an extra blanket in winter; they should be applied over padding (gamgee tissue) as shown in the drawings. It is a good idea to bandage a horse's legs after hard work to support them and prevent swelling. Exercise bandages of stockinette or elasticated crêpe, also over padding, help protect the legs against knocks and jars, especially when jumping, but need to be applied with care to avoid uneven tension or inhibited circulation in the leg. Of course, a bandage which slips or comes undone can cause a nasty accident, and if support or protection against interference injuries, such as brushing or speedy cut, is required, boots are better.

Cleaning and care Leather tack should be washed with a sponge and lukewarm water, then,

Above *Hot water pipes keep the leather in this tack room warm and dry.*

Below *To apply a leg bandage leave a turnover at the top and bandage down over gamgee. The bandage will make a natural turn up at the fetlock and then continue upwards, tying and tucking in.*

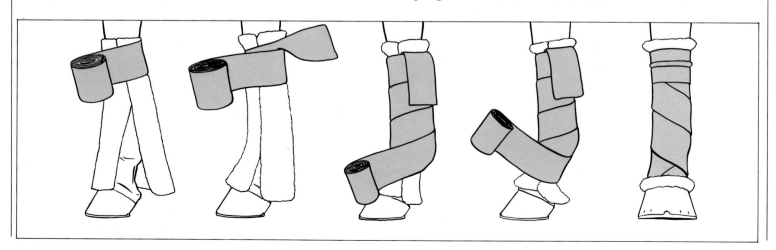

Far left *A top class show jumper illustrating correctly worn tack—a well-fitting snaffle bridle, running martingale, essential stops on the reins, breast plate to prevent the saddle slipping back, skeleton knee caps, open-fronted tendon boots, and a comfortably-fitting general purpose saddle and numnah.*

Left *Exercise bandages are seen here properly applied on a show jumper. They should be put on over padding and for competitive use should be sewn in position to prevent them coming dangerously undone. Exercise bandages normally finish above the fetlock, but a proprietary make of lightweight adhesive bandage can be fitted down over the fetlock.*

using a different sponge (dampened), glycerine saddle soap should be rubbed well into the leather. Metal parts (other than the bit's mouthpiece) can be cleaned with metal polish. Keep an eye on the stitching of your tack and the fit of the saddle, getting any repairs done promptly by your saddler. Regular cleaning and care will keep your saddlery safe to use for many years.

Fabric items can simply be washed in a washing machine, any leather parts being subsequently oiled with a good leather dressing such as neats-foot oil. New tack should always receive a dressing of neats-foot and occasional reapplication will prevent the leather from losing its suppleness, especially if you have been out on a wet day.

Bringing a horse up from grass
The change from a rest at grass to a largely stabled life can be considerable for a horse, and must be done gradually to avoid upsetting him mentally and physically.

For the last few days at grass, start giving ½ lb (225 g) concentrates daily to accustom his digestion to hard food. When you start stabling him, try to reduce the grazing gradually over at least a week, and give mashes and soft hay or hay-age to help his digestion through the change, gradually introducing small, normal feeds after a week.

He should be kept in an airy box with hay and water always available and have walking exercise

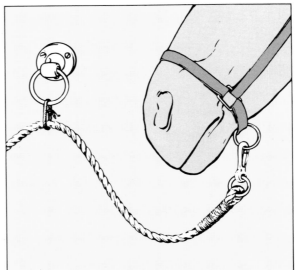

Left *A horse should never be tied directly to a metal ring. In cases of emergency it is better for something to break. Therefore, the halter rope should be tied with a quick release knot to a loop of ordinary string— baler twine does not break—on the ring.*

only for at least two weeks, starting with half an hour a day and building up to two hours at the end of a fortnight.

The farrier will be needed to shoe him, and the veterinary surgeon should check his general health and condition, examine his teeth and advise about a worming programme before you start getting the horse fit for work.

Getting fit
There is only one way a horse can become fit for work, and that is by having his body subjected to increasing amounts of physical effort and stress over a period of several weeks or months. A fitness programme cannot be rushed or the horse could be subjected to strain and be put back in his work. It takes six weeks to get a completely soft, unfit horse half fit and a further six weeks to get him ready for hard hunting, horse trials and the like.

After the first fortnight, introduce short periods of slow trot, which can be increased gradually over the next two weeks. Lungeing sessions also may be beneficial if restricted to a few minutes on each rein and approached with care regarding the horse's condition and state of fitness. The concentrate ration might now be about 3–4 lb (1.5–2 kg) daily, plus *ad lib* hay and water, for a 16hh horse. Take care that the saddle is fitting properly and is not causing sores, especially if the horse has come in fat.

By the fifth week, the horse should be able to trot steadily, but up to his bridle for two periods of ten minutes each exercise session. Cantering can now begin, starting with a couple of minutes a day and building up to five minutes by the end of the sixth week. The horse could now be eating 6 lb (3 kg) of concentrates daily, and is now half fit.

The importance of slow work—smart walking and steady trotting—on hard roads to toughen up his legs cannot be overestimated, especially in the early weeks, if you want your horse to remain sound for several months.

From the seventh week on, the horse's canter work can gradually increase in speed and length of time until by the end of week eight he should be able to canter for ten minutes twice each session, with trotting and plenty of walking in between, and be out for two hours. Feed him about 8 lb (3.5 kg) of concentrates.

Depending on his work, schooling exercises and jumping should have been introduced to jog his memory and develop appropriate muscles, and in the later stages of a fitness programme, cantering should not be done every day. About twice a week, half- and three-quarter-speed gallops should be performed, with an easy day following and interesting, active hacks on the other days. Only you, his owner/rider, can judge now just how much work and food he needs to keep fit and well without going stale.

Regular playtimes in the field will do much to keep him fresh and happy, as will one day a week off with *no* concentrates (to clear his system), but mashes, hay and grazing.

After a hard day
The care your horse receives after a hard day's work will determine how well and quickly he gets over his exertions. Hunting, a one- or two-day event, a long competitive or non-competitive pleasure ride or a long day at a show can all take more out of a horse than he shows.

He should obviously be examined for injuries and treated accordingly. Get him into his box as soon as possible, having first washed him down if he is muddy and thoroughly dried him off, as described earlier. This makes him feel fresher and more comfortable and will not result in skin problems provided he is dried properly.

He needs easily digested, nourishing food, like mashes with cooked grain (have the grain cooked and ready for your return); the addition to his feed of ½–¾ lb (225–340 g) glucose powder (a 'ready digested' sugar) will act as a good pick-up too. Therefore, when he is dried, rugged-up and warm, give him a drink of just warm or 'chilled' water (half a bucket) if he has not already had one, then his hay and mash, and leave him in peace while you see to yourself.

Return later and offer some more water and check him for warmth or breaking-out in a sweat from tiredness or exciting memories. If he has been well rubbed down and settled with his feed and a good bed, however, this should not be a problem. Later still, when you say goodnight to him, he can have another mash if he has finished the first, or cooked grain in damp bran, plus hay and water.

Next day In the morning, look immediately to see how he seems in himself. Has he eaten up? This is your best guide to how he is feeling. Trot him up to check for lameness and if all is well decide whether you are going to ride him out for a short walk, turn him out in the paddock or lead him out. A period of grazing will certainly help him. Stick to cooked feeds and mashes on this day, returning to normal the day after. Even after a hard day, a day off or on light work should always be accompanied by few, if any, concentrates unless you want to run the risk of azoturia, lymphangitis or other troubles. It is quite wrong to increase the feed to compensate for energy expended, as some people seem to think, and a fit horse will benefit, not suffer, from the break.

If he has sweated much, an electrolyte supplement could be needed (consult your vet) to replace minerals and salts lost in the sweat.

Let him have plenty of rest, grazing either in-hand or, preferably, free. Give him easily digested food, and keep your eye on his appetite. When he is

eating up normally, you can be confident he has recovered.

Letting down and roughing-off

Every horse enjoys a holiday, and most like to spend theirs at grass with their friends. Your circumstances, his work and the time of year will determine when, and for how long, your horse will be off work. The fitter he has been and the harder he has worked, the longer his rest can be. For 'family' horses in work most of the year, the ideal time for a holiday is spring or autumn, as the weather is mild and there are few flies around, although rich grazing must be avoided in spring as it can easily cause laminitis.

Assuming it is spring and you are roughing-off your horse for a few weeks' break, choose a mild day and leave off one of his blankets. Stop body brushing him so his coat will become greasy, giving some protection against the weather, reduce his concentrates by about a quarter, and also the length and speed of his exercise. Introduce grazing with half an hour a day for a week, or increase his time out if he is on the combined system.

After a few days, leave off another blanket or rug, cut down the concentrates to half the normal ration (making sure he has plenty of hay) and give walking exercise only. Also increase grazing.

During the second week, provided the weather is mild enough, try to get him to do without clothing during the day, but rug him up at night. Leave him out for several hours, but if he genuinely seems

Above Horses not only enjoy a holiday out at grass, but also need the rest from work at least once a year. The late spring when the weather is mild and the grass is new is ideal. A previously stabled horse needs to have gradually extended periods in the field and his concentrate rations gradually reduced before he can be turned-out completely.

cold, put a New Zealand rug on him. Concentrates should now be down to a quarter of his normal ration and you could stop exercising him if you wish.

If there have been no setbacks in the weather, by the end of the second week he should only need a rug at night, if at all, and be out all day, albeit with a New Zealand rug if it is chilly, early spring weather. It is a good idea to keep feeding him a few pounds of concentrates morning and night.

By the end of the third week, you should be able to leave him out all the time, if that is your plan, but watch his coat growth and, if needed, leave the New Zealand rug on him. A concentrate feed a day in the field for a week, or until the grass makes a good showing, will help him acclimatize all the better. Have his shoes removed and his feet trimmed.

When roughing-off in the autumn, you may well need to carry on with the feeding and the New Zealand rug, despite the growth of his winter coat, because of the deteriorating weather and lack of nourishment in the grass, particularly in late autumn.

At any time of year, watch the horse carefully to see that he is not too cold, or standing miserably by the gate wanting to come in or—just as bad—losing weight. A 'holiday' under these conditions will do him no good at all and if you cannot put things right in the field, the horse might as well come in again. Acclimatization takes time and perception, and, like a fitness programme, cannot be rushed.

Keeping a horse at grass

The mental picture conjured up by the phrase 'out at grass' is that of a relaxed, happy horse, idling his days away contentedly munching grass with never a thought of work. This ideal state of affairs does sometimes exist, but very often things are quite different. By no means all horses would be happy if kept permanently at grass, because, although the horse can eat and exercise freely and associate naturally with his companions, life at grass can be very stressful, mainly because of the weather in winter and the heat and flies in summer.

A horse can certainly be worked actively from grass, with care, but he will never be quite as fit as he would be if at least partly stabled. He will very likely be wet and muddy or dry and dusty when you go to fetch him and you will have to clean him up before you can work him.

The only way you can assess whether your horse is a suitable candidate for a life in the open is to study his behaviour and condition when out. If he spends much time by the gate, if he greets you frantically whenever you arrive, if he seems generally downcast and unhappy and if he is difficult to keep weight on despite regular worming and adequate food, he would be better partly stabled.

Choice of field

The subject of field safety is mentioned earlier; security is another feature worth considering in these days of increased horse stealing for the meat market, and the very best way of deterring a thief

Left Shelter and amiable company are two important factors to consider for horses living at grass. A sturdy three-sided building with a wide opening facing south or away from prevailing winds allows the animals to shelter from extreme winter weather and to escape from the discomfort of heat and flies in summer.

so a field with poorer grass is better, but a field full of weeds is no good either and fields should be inspected for poisonous plants, e.g. ragwort, yew, laurel and laburnum. On slightly poorer pasture you can safely feed corn and keep the horse fitter for work. With today's pollution problems, it is probably better if he cannot reach any ponds and streams, but has water either from a piped supply to a trough or from some other large, safe container filled by hosepipe. The field should be inspected regularly for rubbish.

The importance of shelter

Shelter is an item that is overlooked or disregarded by some owners, who appear to think that as a horse is a creature of the great outdoors, he can put up quite happily with biting winds, driving rain, hail and snow, scorching heat and flies. But domestic horses at grass are denied the opportunities of wild animals to seek their own shelter over several miles, and *must* have really effective shelter if they are not to suffer physically and mentally, contract exposure ailments and go down badly in condition.

A strong field shed is essential for horses who are out all the time. It can be constructed cheaply by any handyman out of used timber, provided it is strong, gives about 2 feet (60 cm) headroom or more and is big enough for all the field's occupants to move around in. Site it on the driest part of the field with its back to the prevailing wind or, if there is no significantly prevailing wind, facing south. It should be three-sided with the front open to encourage easy entry and exit. Keep it well bedded down on deep litter, well stocked with hay in winter, and the horses will gladly treat it as a welcome haven of rest and comfort all year round.

Shelter belts of trees and high hedges (thick all the way down) are also highly desirable.

Company and herd hierarchy

Most horses need company and will pine without it, but horses living permanently together in freedom must be good companions. Although every herd has its leader, its doormat and all ranks in between, any serious troublemaker must be removed for the good of the rest.

A new herd member must be introduced to one existing member separately to ensure he has at least one friend before he meets the others. Then he should be exercised or led in hand with the herd. Simply to put him in with a herd and hope for the best is asking for a badly kicked horse.

Grassland management

Grass is at its most nutritious in spring. Nutrient levels slacken off as the grass coarsens in summer, but there is a moderate flush of growth in autumn. During the winter months the feed value can be disregarded, and supplementary feeding will be essential.

Above The day after a hard day's hunting or competition, the horse should be trotted up in hand to check its soundness.

from taking *your* horse is to have it freeze-marked or branded.

Gates should be kept padlocked at both ends (preferably top and bottom at each end) to make it more difficult for them to be lifted off the hinges. Wooden or wire fencing is certainly no deterrent to thieves, but thick, prickly hedging is, as are houses nearby with occupants who will keep an eye open for suspicious characters.

From your point of view, you will need a field within reasonably easy reach, and for the good of your horse, it will need to be on well-drained land, not too rough or hilly and fairly well sheltered. A horse kept on its own will need at least 2–3 acres (1–1.5 ha) and for each additional horse add at least another acre. Rich grass, such as that sown for dairy cattle, causes serious health problems such as laminitis, azoturia, lymphangitis and colic in horses,

Horses are choosy grazing animals. The parts of a field they like they will crop down to the soil, leaving other 'lavatory' areas, where they do their droppings, quite untouched. Leaving horses on a field, never resting or treating it, will ruin it very quickly, so valuable grazing is lost.

If possible, divide your land into three, or at least two, sections (perhaps with electric fencing which can be moved to allow access to shed and water whichever part is in use) and use one at a time in rotation. When the grass begins to look patchy, move the horses on and try to get cattle on the first field to eat down the long grass. Mow what is left, harrow the field both ways and rest it, and do the same for each section.

Grass is a crop which will only respond with care and attention. Soil analysis by a fertilizer firm, together with advice on treatment, is usually free and the subsequent application of the recommended products will save you money on supplementary feeding. Cattle manure is particularly good for horse paddocks when you can get it.

Picking up horse droppings as often as you can, and worming the horses every six weeks, will keep a firm check on parasite infestation, but *every* animal in the field must be included, and should be stabled for a day after worming to ensure the pasture is not immediately reinfected. Take advice from your vet on suitable wormers.

Flies
The worst part of summer for horses at grass is the constant, tortuous attention of irritating and biting insects. Not only do they drive horses frantic and cause them to jar their legs on hard ground in trying to escape, but they also seriously irritate the eyes, infect wounds and inflict painful bites (according to species). Bot flies lay their eggs on horses' legs, appearing in clusters of small yellow, pollen-like ovals; the horses lick them off and the larvae hatch inside, attaching themselves to the stomach lining and causing damage and colic.

The only fly repellants of any real use are those marked 'residual', available from agricultural suppliers and some saddlers. Their effects last for several days and they have a build-up effect. Start using them before flies become numerous, and follow the instructions carefully, to obtain maximum effect throughout the summer.

Exposure ailments
Constant exposure to wet weakens the skin and makes infection by bacteria quite common, causing mud fever (on the legs) and rain rash or scald (on the back). In both conditions, the skin becomes hot, sore and swollen, with eventual cracking and scabbing, and infected pus underneath. They can put a horse out of action very quickly and take a long time to heal, so watch constantly for the first signs, and if they appear bring the horse in and call

the vet immediately, as neither is a subject for home doctoring. Watch also for the first signs of sweet itch, an allergic response in some ponies to a particular midge which causes them to rub their manes and tails violently. The midge is prevalent in early and late summer at dawn and dusk and if the animal can be stabled at these times, the condition may be avoided.

New Zealand rugs
The comfort a well-fitting, properly maintained New Zealand rug can bring a horse can save several

Below A horse used to being kept at grass will live out quite happily all winter if provided with a well-fitting New Zealand rug. The type without a surcingle and fastened with hind-leg straps is preferable as it eliminates pressure on the horse's sensitive spine.

pounds of feed a day in winter. Keeping him dry and shielded from the wind and cold lessens his concentrate requirement.

The best type is the kind without a surcingle, which is kept in place by shaping at shoulder and hip, and by hind-leg straps.

For a horse permanently out, you will need two rugs so each can be dried off and brushed daily. New Zealand rugs are made in either synthetic fabric or canvas, lined with wool (synthetic fabric is lighter and easier to care for). The rugs must be washed periodically (if synthetic) or scrubbed down outside and vacuumed inside (if canvas), and then re-waterproofed with tent reproofer. Leg straps, if leather, must be kept clean and oiled to keep them soft.

New Zealand rugs are particularly valuable for trace-clipped animals living out and working, or for fine-skinned sensitive animals. They *must*, however, be removed twice daily, the horse checked for rubs and the rug replaced, or changed if it is very wet and dirty.

Feeding

During the grazing season supplement feeds are not necessary unless the field is overstocked or the grass browns off in very dry periods, but horses living—and maybe working—off grass will need extra food from autumn to spring, as has already been explained. As much good hay as they will eat (fed in nets or a long rack in the shelter) plus concentrates sufficient to keep their body weights up are required. A horse at grass in winter, particularly if trace clipped and working, will need even more food than a stabled one to keep out the cold, so the best hay should be used to obtain maximum feed value from it. If the hay is really good, you might manage with two feeds of concentrates daily. Miss out the bran and give 4 lb (2 kg), or at most 5 lb (2.5 kg), in each feed of bruised oats and/or barley with a double handful of chop or chaff (hay and/or straw chopped up small) to aid digestion. Flaked maize can be used if loss of weight is a worry, and the condition of skin and coat will be improved by giving boiled linseed in the feeds two or three times a week. Another useful feed for horses at grass is a proprietary coarse-mix, a mixture of grains and other feeds sold ready mixed by feed merchants on either a national or a local basis. Cubes can be dangerous for grass-kept animals, as they can impact with the grass in the stomach, causing serious colic, or be bolted down by a hungry, cold horse and cause choke.

Do not tip concentrates on the ground where they will be wasted, but feed in bowls, buckets or travelling mangers hooked on the fence. Do stay to see fair play until all horses have finished or the timid ones might have their food stolen by others. If the horses are quiet and good friends, they can be fed together in the shed.

Grooming

Grass-kept horses should only be dandy-brushed, have manes and tails tidied, and faces and docks damp-sponged and dried. They need the grease in their coats to help protect them against the weather. The mane and forelock should not be shorter than about 6 inches (15 cm), which is a compromise between appearance and protection. In winter, the tail should be shortened to just below hock level to prevent clogging by mud or snow, and the heels left untrimmed for drainage. Unpulled tails are probably better on grass-kept horses— they can always be plaited for special events. The hairs beneath the jaw in winter can be trimmed too, to improve appearance.

Only a low trace clip should be considered for a grass-kept horse in winter, which will allow him, with a New Zealand rug, to live out and work and, if trimmed as above, look reasonably presentable as well.

Above *Fields and paddocks must always be checked for poisonous plants and the pasture kept clear of droppings and weed-free.*

Far left *Although these horses look 'run up' or thin, they will cope with snow-covered ground. However, throughout the winter they should be provided with hay twice a day and offered some concentrated feed, moistened with water. It is also essential to make sure that the water trough is kept free of ice.*

3

VETERINARY CARE

The practical equestrian seldom thinks of his horse in terms of scientific labels: in zoological terms, '*equus*' is a mammal, a gregarious, nomadic herbivore and a soliped ungulate.

The meaning of such esoteric tags, however, does have an effect upon the domestication and welfare of the horse and it is in his best interests that you understand the implications of his natural herd instinct, his inclination and need always to have food passing through his system and his basic yet intricate structure, which is, fortuitously, suitable to carry a rider swiftly and surely on his single-hooved feet.

Only by developing your knowledge of the horse in general, and your powers of perception for your horse in particular, will you learn what makes him tick. Some people are born horsemasters, others have to acquire the skills, but pity the horse who has ignorance thrust upon him.

Natural instincts
The natural herd instinct is perhaps the most important aspect of the horse's psychological make-up that the owner/horsemaster should understand, especially if the animal is to be kept at home. Horses do not appreciate isolation from others of their kind, to a greater or lesser degree. Some become desperate for company, possibly escaping from their isolation in field or box, and become a menace to ride. This characteristic behaviour is thus overlooked at man's peril and should be a prime consideration in where to keep a horse.

Although the herd instinct is still strong and must be acknowledged for the well-being of our horses, man's domestication of the horse has taken him out of his natural environment and, in so doing, has created for the horse problems connected directly or indirectly with this domestication.

The greatest set of problems has come with man's attempts to improve the horse through selective breeding over the past 300 years or so. Through selected foundation stock he has bred hardier, stronger and faster animals which have been brought to a more productive performance.

However, the greatest incidence of acquired defects occurs when the horse is brought into the rigours of work and an 'unnatural' environment created by man: the need to shoe brings its own set of problems, while it is interesting to consider that wild horses are comparatively free of worm larvae, and only the stabled horse needs his teeth rasped. These defects include the effects of injury, illness and disease and it is sobering to consider that the majority of riding horses are permanently affected to a greater or lesser degree. Technically they are all 'unsound', but not necessarily unfit for work. They may not be as useful as their potential suggested but, in each case, if deemed fit for some degree of work, a horse will always benefit from good horsemastership.

Horsemastership at a primary level is not a matter of knowing *what* is wrong with your horse but that *something* is wrong. Given that most horses are technically unsound, you should build up a picture of what is normal and constant for your horse, and the mental check list of normality should be automatic on each visit to your horse. His Appearance, Behaviour and Condition should be observed as signs emanating from the horse's various body systems, their structure and functioning, and recognition of something wrong should be immediate relative to the severity of the condition. The potential of the athletic working horse is a reflection of his normal condition but a horse is only ever as good as his weakest part. If he has good legs but a bad heart, is sound in wind, heart and eyes but lame, his potential is diminished.

Experience gives the horsemaster an understanding of defects and disease and the knowledge to act correctly in terms of first aid, calling for professional help and successfully co-operating with the veterinary surgeon in managing and nursing the sick horse. To understand your horse better a knowledge of his ordinary functions and systems is essential. By understanding how his body works, it is easier not only to appreciate the problems when he is affected by illness or injury, but it will also make you a better, thinking rider.

The Respiratory System

Respiration—breathing—is the exchange of atmospheric oxygen for cellular produced exhaust, carbon dioxide. Without oxygen living tissue cannot survive.

When the horse breathes, air is drawn down into the lungs, from where the oxygen is transported across the cell walls, through the cells of the alveoli—small air sacs in the lungs—and into the blood capillaries, thence to the heart and so throughout the body. Conversely, carbon dioxide, the waste product of respiration, is lost from the capillaries to the airways and out of the body. To the horsemaster it is simply the breathing in and breathing out, its rate and depth and the associated sights and sounds that matter. He perceives them as a guide to normality and as a measure of fitness.

Nature conserves energy so, at rest, the horse breathes slowly and shallowly to obtain sufficient air for basic living. Such breathing is silent unless listened to through a stethoscope over the chest wall when 'in' can be heard going through the bronchi and bronchioles but is silent on 'out'. The expanding chest is effected by a backward movement of the diaphragm—a strong band of muscle—which compresses the abdominal contents and so forces the belly wall outwards, backwards and downwards: this movement can be seen, counted and its excursion gauged. There is an inspired hold but no interval between expiration and inspiration.

If the horse moves, becomes excited or frightened, there is an immediate increase in the speed and depth of his breathing and a loss of hold. As the horse breathes faster a noise, heard only when breathing out, is produced as the compressed air escapes into the atmosphere. In-breathing is still silent to the ear. A cough is produced by a rapid forced expiration from the mouth, but if directed down the nose it is a sneeze.

As the body-heated air meets a colder atmosphere it condenses to form the characteristic puffs

Overleaf The healthy head of a fine young horse.

Below An anatomical illustration showing the respiratory system in detail.

The respiratory system

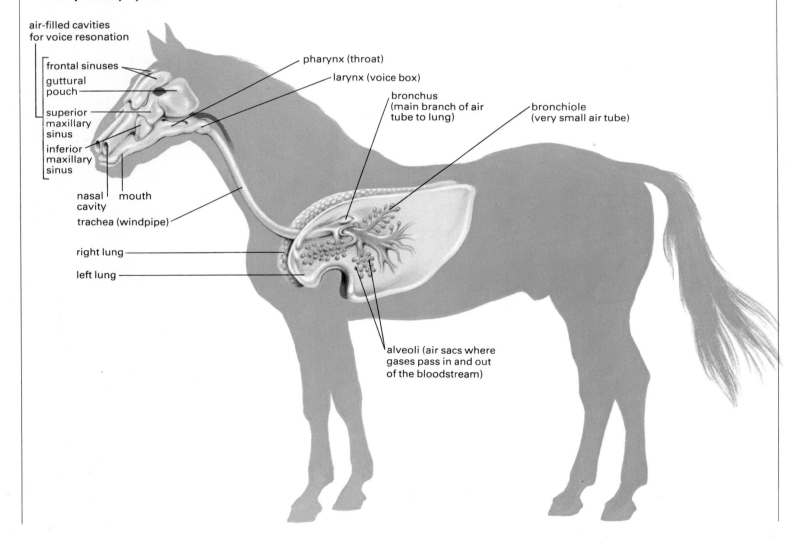

air-filled cavities for voice resonation

frontal sinuses
guttural pouch
superior maxillary sinus
inferior maxillary sinus
nasal cavity
mouth
trachea (windpipe)
right lung
left lung

pharynx (throat)
larynx (voice box)
bronchus (main branch of air tube to lung)
bronchiole (very small air tube)
alveoli (air sacs where gases pass in and out of the bloodstream)

at the nostrils. At faster paces, breathing locks on to the triple rhythms of canter, for example, so that this visible breathing out coincides with foreleg strike, and breathing in with suspension. After exercise, breathing, visible by the horse's heaving flanks, continues fast as the body repays the oxygen debt incurred by severe muscular effort. The time taken for it to return to the resting rate—8 to 15 breaths per minute—is a measure of fitness. This to-and-fro fast movement of post-exercise air also helps to dissipate the heat generated by work.

To meet the demands of work parts of the respiratory system dilate to facilitate air movement. The horse flares his nostrils when breathing out and the head and neck are extended to straighten the 'pipeline', and, unseen, the larynx—the voice-box—opens. To produce voice, the larynx is partially closed so that the vocal cords are tensed and vibrate in the outstream of air. The noise of 'warning' is produced by snorting down the nostrils.

The horse depends on its ability rapidly to increase the volume and rate of air flow over the larynx and the throat to expel threatening foreign material and so does not as often require the safety factor of the cough as does man.

Respiratory problems

In the respiratory system a to-and-fro situation exists, and this keeps any inspired germs on the move, in and of course out: they don't get time to settle, colonize and invade.

By practised observation you should learn to recognize alterations in breathing—rate, depth, noise. Fever will speed up metabolism—the rate at which the body burns oxygen—and generate a need for more oxygen, resulting in faster, deeper respirations and increased loss of body heat: the horse's normal temperature is 100.5°F (38°C) taken by placing a thermometer into the rectum and holding it there for one or two minutes.

The fever need not arise in the respiratory system, but when it does, additional signs appear— the rate and depth of breathing increase and expiration is often forced with a double lift of the abdominal wall. Inspiration is often audible and there is a need for periodic emptying of the lungs so a sudden explosive expiration is involved; the horse coughs.

Infection is often associated with discharge at eyes and nostrils and signs of a sore throat appear, causing difficulty in swallowing and a soft cough. Obstructions to the inward air flow will invariably produce a noise e.g. the horse with a hemi-paralysed larnyx will become a whistler or a roarer which will be more readily heard at faster paces and with the head flexed on the neck.

Prolonged heaving flanks after exercise is a warning that the muscles have been overstressed relative to the degree of fitness.

The digestive system

A horse can be seen to eat and to drink and to defaecate—heard to eat and chew and to swallow and, if listened for, heard to have 'tummy rumbles'. What goes on inside is of course hidden but the results and benefits of eating, digesting and metabolizing food can be assessed by a consistent body condition relative to the work done.

The herbivorous characteristic of the horse requires certain specific features. Grasping growing grass, oats or hay requires manipulative and powerful lips which are the horse's 'fingers'. They are the most mobile and sensitive part of the horse who thereby can separate out unwanted from desired foodstuffs and recognize injurious non-foodstuffs. (Somewhat unfairly man makes use of this sensitive proboscis as a point of restraint to which a twitch may be carefully applied.)

Herbage gathered by the horse's lips is cut by his 12 incisor teeth and his powerful tongue draws it back between the 24 rough-tabled molars which grind the food into small, lacerated pieces.

Chewing is a very necessary function and leads to three factors in the digestive process. Salivation occurs only by stimulation of chewing, and although saliva contains no digestive enzymes, it does have chemicals which begin the acid-alkali shifts found down through the stomach and intestines. Without it digestion gets off to a poor start. The tearing action of chewing produces many surfaces of grass stem and leaf. These pass almost unaltered to the blind gut, or caecum, and to the four compartments of the large intestine. Here the cellulose constituent is acted upon by specific micro-organisms to produce volatile fatty acids, the horse's main source of energy. Thirdly, the action of chewing releases the soluble sugars, proteins, minerals and vitamins of the herbage which become available for 'ordinary' digestion in the stomach and small gut, or further on as energy-rich material for the micro-organisms' multiplication.

Significantly, the species of micro-organism suited to a particular crop differs from others. All are present—originally by the suckled foal eating his dam's faeces—but need 7–10 days to become fully recruited. Therefore sudden changes in food-stuffs must be avoided. Normal eating can be judged by the horse either being seen to graze regularly or to clean his manger and empty his hay net (but not in either case onto the floor!).

If kept at grass the wear and tear on the horse's continually growing molars rarely causes any irregular growths but with hay, straw and cereals, all of which contain some lignin and silicates from the soil, the wear does produce points and sharp edges on the outer side of the upper molars and the inside of the lower. Sharp and therefore painful teeth are often overlooked as the cause of bit evasion or a horse not eating up his feeds.

Digestive disorders

The horse eats contaminated grass and drinks unsterile water; it licks its and others' coats and so many organisms are swallowed to pass along, through and out of the digestive tract. Many multiply within and nature repopulates the environment via the faeces. Many are necessary inhabitants of the gut, the large intestine especially, for the micro-organismic breakdown of cellulose (plant carbohydrate). Some are potentially pathogenic, and in the first instance they are contained by physical defences. Gut content flow does not permit time for colonization; changing acidity gives chemical defences and an inner skin, the endothelium or mucus membrane which lines the gut (which is but an internal surface of the body) is a further line of defence. A very complicated system of mucosal cell and blood bone cell are constantly present in the deeper layers to defend penetration.

Signs of abnormality in the digestive system include not eating, inability to eat, or eating with difficulty; inability to swallow, or with difficulty, quidding, i.e. expulsion of food from the mouth;

evidence of abdominal pain (colic); droppings of unexpected colour, consistency, smell and amount; and failure to maintain body weight.

Thus a sick horse goes off its food. One with sharp points and edges to its molar teeth begins to eat messily and eventually will quid out mouthfuls of partly chewed food.

Incorrectly fed concentrates, or food taken whilst still in post-exercise fast-breathing, will occasionally become stuck in the gullet causing choking with dramatic signs of distress.

The many causes of colic are too numerous for this chapter. However, the horse evinces signs of severe pain when certain defects develop in the digestive system. Briefly, the common types are due to impaction of improperly digested food somewhere along the small or large intestinal tract. In impaction, the lumen diameter of the intestine varies, sometimes quite markedly, thus causing a bottleneck to coarse or fibrous foodstuff. There is a gradual build-up until normal fluid contents are dammed back behind the blockage, causing moderate but worsening pain. Eventually changes occur

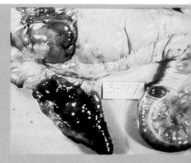

Above *A photograph of* Strongylus Vulgaris, *the parasitic redworm which produces severe effects in horses.*

Below *The horse's digestive system.*

The digestive and excretory systems

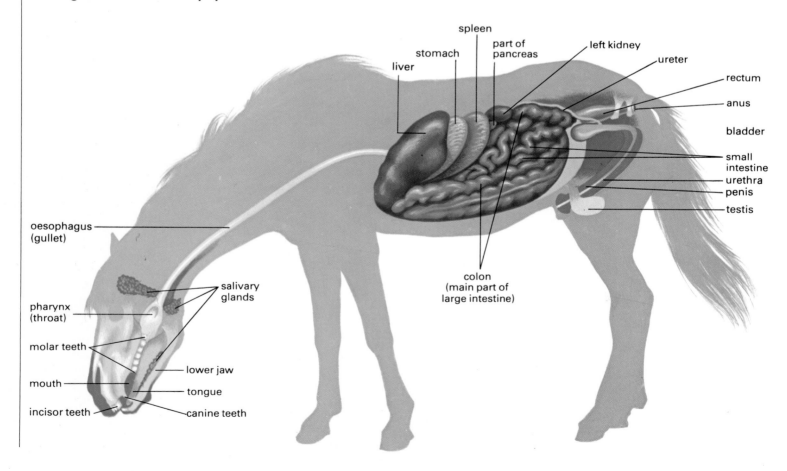

spleen
part of pancreas
stomach
left kidney
liver
ureter
rectum
anus
bladder
small intestine
urethra
penis
testis
colon (main part of large intestine)
oesophagus (gullet)
salivary glands
pharynx (throat)
molar teeth
lower jaw
mouth
tongue
incisor teeth
canine teeth

Above *Bots in the stomach which will cause pain and discomfort.*

in blood fluid chemistry with more serious consequences which could lead to death from electrolyte imbalance and damage to vital organs.

A horse will show the characteristic signs of colic by pawing the ground with a foreleg, lifting a hindleg towards the abdomen, and by staring round at his body with an anxious expression and obliquely laid-back ears.

The pulse increases in speed but rarely above 60–70 per minute. Some patchy sweating may occur. Little or no faeces are passed after the first 24 hours during which the only signs are a reduced appetite and a lack of sparkle. Thereafter attempts at defaecating and urinating alternate with intermittent picking at food and lying down. The horse may roll but, especially in large intestinal and blind gut (caecal) cases, he may prefer to lie half over on his back with legs propped against a convenient wall.

The more dramatic colics are associated with obstruction to the blood supply to the large intestine and caecum. Intense cramp pains occur with violent rolling, sweating and rapidly rising pulse. When this passes the 70 mark it is, to the veterinary surgeon, a warning of impending crisis usually associated with gangrene of a length of bowel or a twist leading to gangrene. Exploratory surgery becomes essential, and modern developments in anaesthesia and monitoring of electrolyte levels permit this to be carried out with success.

These vascular colics are invariably due to formation of a blood clot in the vessels from irritation caused by migrating larvae of the large red worm (*Strongylus vulgaris*).

Less serious but often dramatic colics occur from simpler digestive upsets—spasmodic colic—or to out of tune contractions of peristalsis whereby lengths of the gut go into spasm. The pain, often intense, is usually intermittent. The cause is an interference with the autonomic nerve supply to the bowel's smooth musculature. This can arise from toxic irritation and reflex hyperactivity. It is often merely described as acute indigestion as such cases can often be related to mistakes in quantity and quality of food given. The standard routine treatment is control of pain. Injectable analgesics are much more efficient than those given as a drench.

The heart and blood vessels

All systems work as economically as possible to conserve energy. At rest the rate of the heart beat slows to just sufficient to maintain basal activity in terms of circulating oxygen, nutrients to and products, hormonal and 'waste', from the organs and tissues.

The heart can be felt to beat by a hand over the ribs behind the left elbow. The undulating flow of blood through the arteries can be detected as the pulse when a medium-sized artery passes over a firm area but just under the skin. A horse at rest has a pulse of 36–42 beats per minute and this is most easily taken by pressure of the finger tips on the artery which passes under the jaw. (Feel it on your own face first and you will easily locate it on your horse.)

With excitement, fear and of course work, the rate and power increase so that the pulse rate goes up and the beat of the heart can be felt by the rider's legs. Once again the time taken to return to resting normal is an indication of fitness and is of greater significance than a similar measuring of respiration. Research has shown that the horse undergoing fittening work must have his heart rate taken to 200 per minute. Even if pulse is taken immediately after galloping this will have fallen to around the 100 mark and soon afterwards slow towards resting rate. As a guide to response to such work the time to return to 50–60 beats per minute is assessed. When no more than 20 minutes, the trainer can progress to more stressing work.

In the horse, the areas of the body surface where the blood colour can be judged are restricted to the membranes of gums and the white of the eyes, but any attempt to quantify anaemia by visual inspection is of doubtful value because of the normal yellowish colour of these surfaces. In the clipped horse or one with summer coat the vast network of underskin blood vessels can be seen after hard work as the result of residual increased blood pressure. This maximal surface volume of heated blood facilitates thermal control through the mechanisms of radiation, convection and of course sweating. Another example of dual-purpose systems.

The skin

In the horse, of course, this protective barrier is seen mainly with hair but it can be felt to be soft, elastic and mobile to degrees which vary with position on the body. The temperature of the skin also varies, and when body temperature rises the skin acts by sweating—the evaporation of sweat helps to cool the body down. A horse's hooves are a modified form of skin which grows as hard horn. Just as the skin itself is constantly dying and being replaced, so the horn of the hoof grows and must of course be maintained through skilful trimming by a farrier.

The quality of the horn cannot be altered once it appears as the wall or sole of the foot , which grows at about one inch every two months. Oiling affects evaporation and therefore helps prevent the hoof deteriorating, but it cannot improve it more than cosmetically. Counter irritants to the coronary band merely stimulate a fast rate of growth of similar quality. This can be altered by feeding additives.

A finger on the pulse

As far as the heart and blood vessels are concerned, the horsemaster must be aware of the pulse rate. This of course increases with exercise, but it will also race in colic, other painful cases and in fevers. Where there is local inflammation, i.e. there is need for more than the usual quantity of blood per unit of time, this is obtained by dilation of the supplying blood vessel whose pulse will feel bigger and fuller but not faster.

Sometimes germs win through the body's natural external defences to invade the cells where they multiply and disrupt the cells' function. Thereafter they gain entrance to the surrounding tissues and into the vascular systems, the lymphatics and the veins from where they may affect the whole body. Following the septicaemia (bacteria), toxaemia (products of bacterial multiplication) or the viraemia (viruses) the particular germs 'settle out' in specific tissues or, in the case of toxins, become widespread. A special case is seen in the toxin produced by multiplication of the tetanus bacterium (*Clostridium tetani*) in which the toxin travels to its predilection seat, the brain, via the nerves from the infected area.

The peculiar arrangement of the blood supply in the hoof causes marked pain when these vessels are congested as in laminitis. Some seven square feet of capillary network swells within an unyielding hoof casing and the pain is intense. If the condition does not resolve within three days, separation of sensitive from insensitive laminae occurs and the pedal bone rotates, with serious consequences.

Lymphatic vessels, an auxiliary return flow for blood fluid—lymph—can show occasional distension as in 'lymphangitis', when an accompanying damming back of fluid in the distal limb(s) causes a soft oedematous swelling—filled legs. The sporadic acute type, Monday morning leg, is a very painful crippling disease associated with fever and malaise.

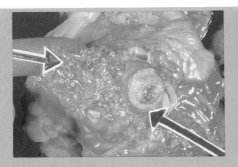

Above *On the left a diseased bone while the photograph on the right shows an infected leg.*

Below *The location of the horse's heart and major arteries.*

The main arteries of the circulatory system

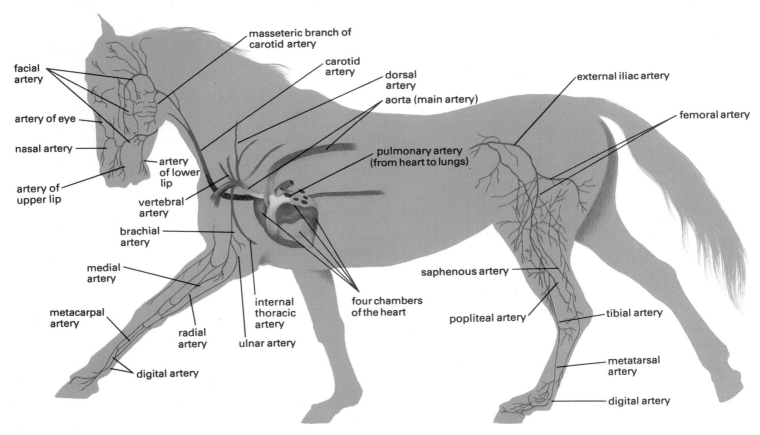

Right A horse at full stretch requiring all its muscular control and poise while on a stiff eventing course.

The musculo-skeletal system

The original country over which the horse roamed in the wild was often hard and flinty. Through evolution the Eohippus genus' toes became one: and that of bone, ligament, cartilage, blood vessels, nerves and horny hoof became a highly specialized organ for locomotion. Man-made roads for the domesticated horse were often too hard and always too slippy for the unshod foot and so shoeing, bringing a new set of problems, became necessary.

To the average horseman the effectiveness of the locomotor system is all-important. The horse is normally used as an athlete and so ability in speed, endurance and agility and durability in its legs are vital. Is he sound in action?

The horsemaster should familiarize himself with the appearance of the gaits of the horse and the normal action of an individual horse within these gaits is again a case for observation and 'feel'. At all times the rider should be able to feel exactly what is happening underneath him.

The farrier will assess leg flight and foot fall at the walk and trim the hoof and shoe accordingly. The horsemaster will have the horse 'trotted up' in hand on a hard, smooth surface to assess levelness and the horseman will feel through his hands and seat the balance of action in walk, trot and canter in straight lines and on both circles. If all is well, work can continue.

A horse's action is geared to the speed required and is set to a certain pace, balance and 'lead' at all times as economically as possible. Work may stress him, but stress leads to improved fitness of muscles if judiciously applied through training. Working out a suitable and correct training programme is of course vital for the horse's well being. Do make sure that professional advice is taken.

No foot, no horse

A lame horse is of little use to anyone. Lameness is an observed alteration from the normal way of going of the limbs, for a particular horse. Most commonly, lameness results from an attempt to minimize pain, but it can also be due to a mechanical interference to movement. The pain in a limb, wherever and from whatever cause, becomes more intense when weight is put on that limb and/or it is made to move. Concussion, compression and stretch can all influence the defect and the horse therefore tries to put less weight on that limb, or puts weight on it for less time or moves it stiffly and off normal line. Invariably this requires an alteration in the centre of gravity and the horse compensates by lifting his head as the affected (fore) limb comes into action. Hind limb lameness may show as a dropping off when weight is taken, or by a raising of the quarter in either progression or weight bearing.

The athletic performance expected of a horse gives rise to occupational hazards. Trauma is the comprehensive term and this may result from contact with hard or sharp objects, from strain exerted within, usually from false strides, or vascular obstructions resulting from concussion and compression.

Traumatic disease

Experienced horsemen become aware of those diseases and injuries which are most likely to affect horses doing particular work. Whether this is show jumping, eventing, dressage or long distance, the horse is athletically employed: he is doing work! The species' locomotor system evolved to enable the wild ancestor to escape by flight. Short distances of fast, jinking speed were followed by longer distances of slower, steady cantering and then an unstressed return to nomadic grazing.

It is accepted that at rest the horse carries 60 per cent of its mass on the forelimbs. At work not only does this weight geometrically increase but, at the faster paces, and especially in jumping, all the weight of horse and rider is taken by the forelegs.

The limbs downwards from elbow in front and stifle in hind perform their support and take-off work as fairly rigid props, while the let-down anti-concussion effect is restricted to fetlock, pastern and hoof joints. These actions demand, amongst other things, tough, resilient flexor tendons and suspensory ligaments and fit extensor and flexor muscles. The hoof itself, with its intricate anti-concussion and auxiliary circulatory pump mechanisms and its stability and durability is subject not only to concussion but also to compression and twist stresses.

The components of the stride which influence trauma are hoof contact, weight bearing and 'take-off'. The elegant 'strain-free' phases depend upon agility, sure-footedness and fitness related to the

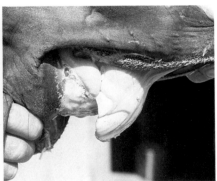

going and influenced by the rider. The tired unfit horse is in danger of pulling tendons and ligaments.

Inherited conformation of the lower limb can predispose to excessive strain and concussion and so lead to lameness. Such inherited or congenital defects can also cause interference of one limb by another or insecure stability but in many such cases the horse 'learns' to compensate for these, or corrective shoeing can minimize risk.

It is important to appreciate that acquired conformation defects can arise through neglected hoof trimming and shoeing: this basically permits an alteration in the lateral pastern/hoof axis whereby the support of the movement is applied too far under the foot, there is an increase in leverage stress and a strain results. Conformation also plays a part in high limb lameness but more frequently this merely limits the usefulness of the horse.

It is not surprising that the clinical list of orthopaedic disease is related mainly to the forelimb from the knee (carpus) downwards, and that pedal ostitis, side bone, navicular disease, osteoarthritis of the three lower joints, and tear of flexor tendons are the conditions which cause the greatest loss of use and even total loss.

The severity and the acuteness of a defect will determine the degree of lameness and therefore the awareness of it by the owner/rider. It is important to detect its presence but not necessarily to diagnose the cause. It is still a truism always to search the foot which should be thoroughly cleaned before a methodical search with hoof knife

Above A selection of photographs showing various deformities in horses.

and hammer is made.

The forelimbs act primarily as a prop to defeat the forces of gravity and the head and neck are the means to elevate the forehand up and clear of the ground for the next period of suspension. This is assisted by the recoil of the kinetic energy developed in forelimb tendon and ligament. The hindlegs, through the quarter and loin muscles drive the body by rotating the femur forwards on a fixed hoof and almost rigid limb whereby the pelvis and spine are pushed forward.

Overstress, or strain, to the hindquarters is related to conformation—cow hocks, sickle hocks, weak hocks—whereby the joint, more or less fixed during the support phase and synchronized, or locked on with the stifle, is exposed to undue strain, or suboptimal stability leading to osteoarthritis (spavin) or plantar ligament strain (curb). These could be precipitated even in good legs by a slip or continuing work on tired muscles.

At canter or gallop, the hind limbs have a differing role dependent on the lead. The first to make contact is primarily a support, a prop against falling outwards, the second is the main engine. Should balance be lost then the smooth propulsion is lost and undue or unlevel stress i.e. strain, is placed upon the ilio-sacral joint or joints, and on the lumbo-sacral joint, giving rise to some of the defects commonly included in back problems. In jumping both feet should share contact in time and degree if strain is not to result. Once a horse is unbalanced for any length of time whether from saving a foreleg and not using a back one properly, or evading some painful condition e.g. sharp teeth, the resultant unlevelness influences the whole spine from atlas to croup so that tensions develop and guarding with resultant stiffness ensues.

A severe concussion or strain will show quickly and clearly—the horse is lame! Persistent or repeated sprain to the small ligament may not present obvious signs until the related bone/joint complex becomes involved. Then an osteoarthritic lameness becomes apparent in the cold horse, but disappears, in the early stages, when warmed up. Quite frequently such insidious lameness is felt through the breeches or seen as a subtle loss of action.

The skeletal system

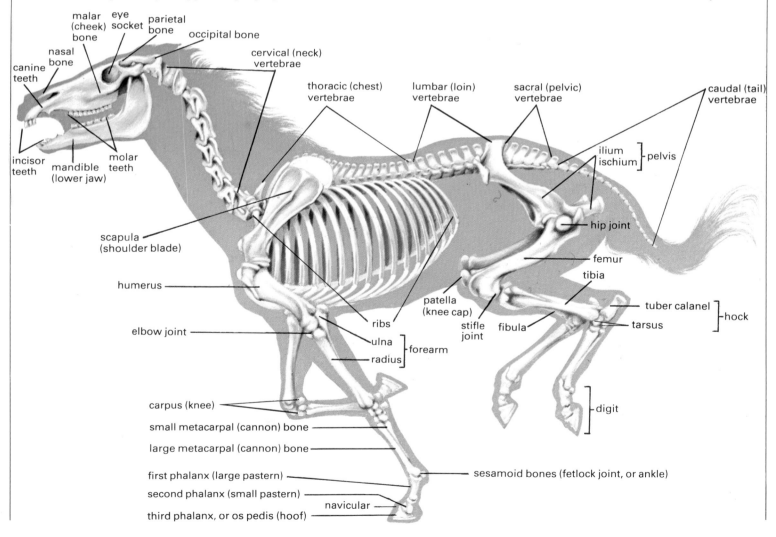

The control mechanisms

The nervous system can be divided into two distinct systems. The sensory system in which the brain receives information about its surroundings by sight, sound, touch etc, and the motor system by which the brain controls the body's actions and reactions. It should be remembered that the horse is primarily a reflex animal. In the wild he hadn't time to think; he responded to stimuli—sound, smell, sight—and reacted appropriately and, if necessary, quickly.

Through schooling the horseman tries to eliminate the reactions of fear and to control the reaction of flight. Training aims to channel and refine the natural reflex to immediate smooth reactions (response to the aids) and to develop the automatic nature of the paces to as strain-free a state as possible.

A good horseman knows how a horse reacts and does not imbue it with man's mentality. He uses reflex activity in the horse to his own advantage. As a simple everyday example, if a horse is reluctant to have a foreleg lifted, have him held with head and neck turned to the opposite side. He will reflexly extend and so prop his 'inside' leg and flex or relax the outside. He has to do so when turning a circle—automatically, reflexly.

More sophisticatedly, a weight on a horse's back makes use of a built-in reflex. The pull of the abdominal contents slung from under the backbone increases with motion; initially the spine gives a little but immediately lifts against this increasing weight and stiffens to cope with locomotion and to give support—support for the horse's body and incidentally, support for the rider. By applying the leg aid behind the girth the rib pressure reflexes cause the pelvis to flex on the end of the lumbar spine, thereby permitting the hind legs to move forward, under and down, and the quarters to engage.

The various bits and gadgets all act by sensory stimulation causing reflex nerve reaction. Good bitting encourages carriage and action by using existing reflexes, although sometimes it is difficult to discern their innate use in the wild. Bad riding through inconsiderate 'hands', implementing aids which work by making the horse consciously assume unnatural positions, may get results through submission to pain and not through automatic responses, but more likely it will cause pulling, yawing, coming above or behind the bit and other evasive action.

The nervous system

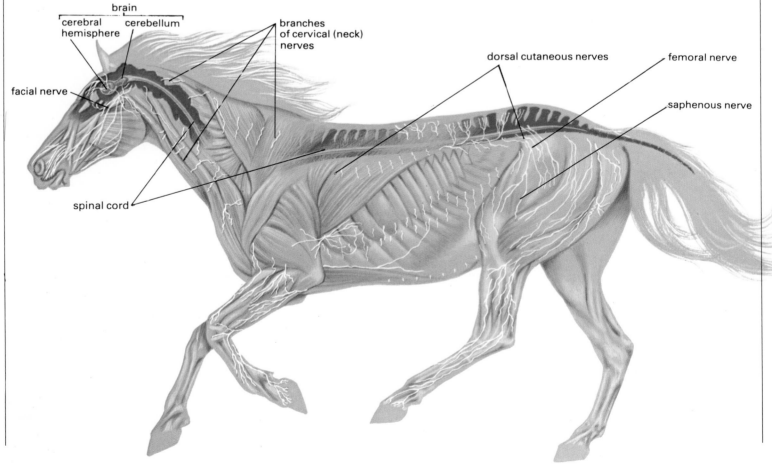

The hormonal system

Nervous control of muscles and body systems enables the brain to give rapid instructions. A slower means of internal communications in the animal involves chemical transmitters. Hormones are secreted into the blood, by specific organs, in which they circulate to reach sensitized or target organs which are stimulated to work, often by secreting other hormones or digestive juices. They play a part in regulating life in ways which need not be hurried or immediate—a biorhythm! They also, of course, play a very important part in sexual behaviour and are the cause of erratic behaviour in some mares during the breeding season.

It is accepted that normality is more readily detectable in a horse 'got up' for work. So too are abnormalities. Certainly the working horse is more likely to become abnormal in health than a resting one and therefore early signs of ill-health in the at-grass animal are more readily missed. The perception of ill-health (dis-ease) is basically to recognize that all is not well, not necessarily to be able to put a name to the problem.

The natural defences

In general terms, metabolic disease is the result of mistakes in feeding and exercising and as a result of domestication. Modern harvested hay and straw and their mould contaminations lead, in some horses, to allergic respiratory disease. Transport increases the incidence of germ infection as unacquainted horses meet at competitions, studs, yards and race courses. Metabolic disease involves no protective response on the part of the horse, and recovery, if it occurs, depends upon the disease so affecting the horse that it stops working and so reduces the stress.

Domestication predisposes inflammatory disease also. For example, asking a horse to jump exposes it to traumatic hazards unlikely to occur at the more natural flat paces, while incorrectly fitted tack can cause sores leading to muscle tension, unbalanced action and more muscle tension.

Non-infectious inflammation is usually the result of trauma without laceration of the protective skin. The trauma may be a single external hard blow, or repetitive concussion when the subcutaneous lesion is commonly known as a bruise. It may be internal in origin, as from overstress resulting in strain of a tendon or sprain of a ligament, but these are still 'bruises' in as much as both concussion and stress result in rupture of subcutaneous and deeper blood vessels with 'internal' bleeding. Inflammation is the process by which:

1 the bleeding is arrested;
2 the debris is cleared up;
3 the area is repaired to a greater or lesser extent.

Understanding disease

Evolution means adaptation to changing circum-stances. As the horse evolved, along with other animals and plants, it learnt to seek out its own kind, to select nutrients and to avoid poisonous plants. It learnt also to live in the presence of micro-organisms which, earlier in prehistory, had adapted to soil-living or to free or parasitic existence.

The horse meets these micro-organisms in everyday life. Some contaminate his coat and skin but these barriers, the dead outer layers of the epidermis and the lifeless hairs, keep the potential germs away from sensitive living tissues. Sweat glands and hair follicles are, under normal conditions also resistant to them; the germs can live there, multiply to some extent but never so rapidly that, by sheer weight of numbers, they invade through the surface defences. Secretions from these areas, in fact, have antigerm properties.

The horsemaster will learn to appreciate the normal texture and elasticity of the skin in different parts of the body. Noticeable changes are slow to develop except in those resulting from inflammation or dehydration, and alteration in the coat, likewise, may take time to become obvious. Nutritional defects will be reflected only in the growth of new hair whereas hair follicle damage produces very quick signs.

Sometimes external physical conditions alter the skin's defensiveness and dermal inflammation occurs as in mud fever, while at other times the follicles are invaded by germs pathogenic in themselves as in ringworm.

If a horse's vitality is lowered, perhaps in winter or through ill-health, skin (endo) parasites such as lice can be propagated causing scratching and in one case anaemia occurs through blood sucking.

Sweet itch, the allergic response to the bite of midges is common and the resultant irritation around the crest and dock can be severe and serious.

If this is seen or suspected in the horse or pony then you must contact your local veterinary specialist as soon as possible in order for it to be treated.

 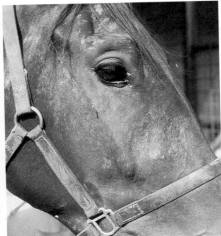

Left The horse's infected eye.

Right A similar infection in this horse.

Combating disease

Living micro-organisms—germs—can, by definition, be killed.

Disinfection implies a removal of germs from an environment: floors, walls, implements, instruments etc, and from body surfaces. Mucking out is a simple but incomplete disinfection of a stable contaminated by germs excreted in the faeces. Cold hosing down, pressure hosing and so on are more effective and steam under pressure, which not only loosens germ-laden organic material but also sweeps them away, are all examples of disinfection.

On clean floors and walls, washed implements etc, the application of disinfectant will almost complete the task. No stable can ever be made sterile!

Skin, hair and horn can all be disinfected but the extent or degree is extremely limited. Ringworm is perhaps the only skin disease that warrants an attempt, not so much by a disinfectant as such but by specific chemical antifungal washes available from your veterinary surgeon.

In other situations where localized infection has occurred, as in a contaminated 'open' wound, the use of crude disinfectants is to be avoided because they kill not only germs but also living tissue. A refined disinfectant, an antiseptic, is much less harmful to eroded skin and mucus membrane. In practice certain specific antibacterial substances, generally known as antibiotics, in the form of lotions, creams, powders, etc, act as antiseptics. As a generalization, however, one of the best cleansing agents for surface wounds is running tap water!

The use of antiseptic agents is sometimes the cause of confusion. The principal aim is physically to remove contamination and to complete the disinfection by applying a suitable substance which, in turn, acts as a barrier to reinfection. Dressings over a wound serve this purpose as well as giving support and physical protection.

Viral contamination of external surface wounds is not a practical problem, nor are fungal infections, except in 'airless', moist situations. It would seem reasonable, therefore, to assume that all surface wounds will be contaminated with drug-sensitive bacteria. It must be remembered, however, that such bacteria are of a mixed or varied type, and not all are susceptible to the same antibiotic. A broad spectrum antibiotic, often mixed with inert antiseptics, is therefore necessary if it is to be effective.

First aid knowledge

If something is wrong, stop, think and if you are sure you know what to do, do it. If not, get more experienced help.

If a foreleg feels swollen and you suspect a leg, make sure that the swelling is not an engorged artery which will fill only the inside of the cannon. Search the foot!

If obviously the flexor tendon(s) is affected, then apply cold water for four or five minutes and pressure bandage. Seek further advice. If the pain is in the foot, clean the sole and frog, including clefts and look carefully for embedded small stones; has the horse been shod recently and the shoe fitted too tightly, or a nail incorrectly placed?

Squeeze the heels, especially if the shoe web appears to have moved off the buttress and dropped onto the seat of a corn.

The farrier's help may be called for in the first instance but remember that, by law, he is not supposed to do more than render first aid by removing the shoe, and advising whether a veterinary surgeon should be consulted. Where insurance cover is held it is imperative that wounds which could incur claims are seen quickly by a veterinary surgeon.

Essential first aid is to clean up wounds: to remove hair from the edges and dead, loose tissue as well as gross contamination from the centre and depth. It is wise not to probe blindly and ignorantly, rather to use the exceptionally good qualities of cold running water. Not only does it flush, or lavage, the area but, after some 10 minutes, it encourages an increased flow of healing blood.

In the horse the tissues from just above the knee and hock downwards do not include muscle (flesh) and so there is no cushion between skin wound and deeper hard bone. As a consequence the in-filling, immature but rapidly growing connective tissue is pushed outwards to mushroom above the level of the wound perimeters: the granulation tissue which prevents 'skinning over'. If, in the early days of healing, the wound has been properly cleaned and infection controlled by topical (local) and systemic (blood-borne) therapy, such lower limb wounds benefit from permanent dressing (up to 10 days) with a porous under-pad and a thick layer of gamgee under firm bandaging.

The horsemaster should know the relatively more dangerous sites of wounds. As examples, over a joint or a tendon sheath as compared with over the side of the cannon bone; down the back of the leg as distinct from over the front; up under the fetlock as compared with over a bulb of the heel. Puncture wounds in these priority areas are more potentially serious than lacerations and grazes. Infection in a sheath or a joint capsule can prove very difficult to control. It is a veterinary surgeon's duty to decide if systemic (oral or injectable) antibiotic is used, as there are dangers in the misuse of such substances.

When a horse is taken out of work immediately stop all concentrate feeding and give a bran mash. Return slowly to full feed, and try to keep it behind the work until that is well established. Following a debilitating illness extra nourishing feed may be necessary but on this your vet will advise. There is nothing like grass (or grass meal judiciously used) and good-quality 'munchy' hay.

The horse whose work is drastically reduced for reasons other than injury or illness—e.g. during frost, snow or rider's illness—must also have a reduction in high-energy feeds. If not there is a grave risk of overloading the muscles with sugar (glycogen) whose burning up during the next hard work will precipitate acute 'azoturia'. Lesser cases clinically, but none the less serious, may come on after work—set fast or tying up.

These metabolic diseases almost certainly have other causes as well and there is evidence that rations overrich in phosphorus, e.g. bran, are culpable. Deficiencies of the trace element silonium especially near the end of winter when vitamin E in the conserved foodstuffs has been used up in 'protecting' the hay or grain may play a part.

Cases vary in severity from being completely rigid behind and even going down, to a mild loss of hind action and tenderness over loins and quarters. All should receive veterinary attention differentially to diagnose from traumatic injury, and to treat accordingly.

First aid cabinet
2 thermometers.
Ordinary kitchen scissors—for cutting cloth, etc.
Half-curved blunt-ended scissors—for cutting hair round wounds.
Gamgee and lint. Cotton wool.
Medicated non-stick dressings (consult V.S.)
10 cm wide bandages.
10 cm wide Elastoplast dressing.
Antiseptic wound powder.
Antiseptic soap.
Antiseptic e.g. Savlon concentrate.
Kaolin poultice or Animalintex.
'Lead' or other cooling lotion.
Emollient ointment (Stockholm tar).

This list will almost certainly increase with time and experience but only liaison with your veterinary surgeon will determine whether you can be entrusted with drugs for oral or injectable use.

Immunization
Use has been made of the body's natural development of immunity or resistance to specific diseases in the artificial protection of man's animals. The virulent factor within a germ is removed or altered (attenuated) and the immunity stimulating factor (antigen) is inoculated to the animal on two or more occasions thereby stimulating the formation of specific antibodies.

In the horse two diseases, tetanus and influenza, are thus controlled.

Tetanus toxoid—an altered toxin—is injected twice the first year, at about six weeks apart, and repeated in one year to be followed at three- to five- year-intervals with booster shots, or should an

obvious wound require high protection. The anti-toxin now circulates, ready to neutralize any toxin produced from the parent germ which has contaminated a wound, usually a puncture or badly bruised one, which, short of oxygen, provides an ideal climate for these soil-living, spore-forming germs to multiply within.

The two recognized strains of influenza are harvested from the artificial medium (living tissue cells) on which they have been multiplying. After suitable laboratory alteration, regulated doses are inoculated but all first boosters and the subsequent shots must by given no further than a year apart.

Some important facts about disease-producing organisms.

Types	Size	Drug susceptibility	Immunity
Virus	Ultramicroscopic	Nil	Recovered—good Vaccinated—reasonable
Bacterium	Microscopic	Good	Recovered—good Vaccinated—variable
Fungus	Microscopic	Reasonable	Recovered—good
Protozoon	Microscopic	Good	Variable
Endoparasite	Adults macroscopic Eggs microscopic	Very good (immature stages not so)	Doubtful
Ectoparasite	Eggs microscopic or macroscopic	Very good	Doubtful

Above Insect bites which lead to great irritation.

Above Sweet itch, an allergic response to midge bites

Above A bad wire cut on the horse's leg

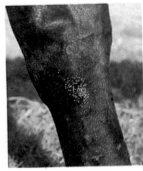
Above Bot eggs on a horse in the autumn

Above A badly infected leg

Popular Breeds of the World

THIS COMPLETE SECTION covers the world's most popular breeds of light horses. Each breed is illustrated, generally from more than one angle. It contains a brief easy-to-read summary of the characteristics of the breed. While some horses are dealt with in great detail, others require less. The range of horses included should give you a thorough grounding in what the best horse would be for you. As well as describing the breeds the entries indicate the purposes to which each breed or type is particularly suited.

The Alter-Real

Height 15hh–15.2hh approximately.
Colour Mostly brown or bay; some chestnuts and greys.
Head Small with a straight or slightly convex profile.
Neck Short, well arched as in the Andalucian.
Shoulders Strong, muscular, well sloped.
Body Short, close-coupled. Deep chested.
Hindquarters Broad, well muscled and powerful.
Limbs Short legs set well under body. Big, flat knees, very strong hocks.
Action Extravagant, with considerable knee flexion resulting in a rather short stride.

The Alter-Real is described as a fine, spirited riding horse with showy action, but with a slightly suspect, highly strung temperament that needs careful handling. On grounds of temperament alone, it is probably not ideally suited to the average rider, but in its native Portugal it has a long and honourable history in *haute école*, where its elevated paces are seen to best advantage.

The first Alter-Real stud was founded in 1748 by the House of Braganza at Ville de Portel, with Andalucian mares from Jerez in Spain, but within eight years it had moved to its present site at Alter, from where the breed takes its name.

The breed has survived a number of crises, including the theft of the best stock by Napoleon's troops, and in 1834 much of the stud's land was confiscated on the abdication of King Miguel. The breeding stock was drastically reduced and the Royal Stud abolished.

The remaining stock was subjected to almost indiscriminate infusions of Hannoverian, Norman and Arabian blood, which caused a rapid decline in the quality and standard of the breed. This was aggravated by further infusions of poor Arab blood. Fortunately, the decline was halted, first by the introduction of the Zapata strain of Andalucians, and then, in 1932, by the Portuguese Ministry of Economy, who took over the stud and who have done much, by culling the unsuitable mares and using only the soundest and fittest stallions, to improve the breed.

That the modern Alter-Real still retains many of its original characteristics says much for the prepotency of its Andalucian progenitors, and the future of the breed now seems assured.

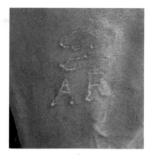

Far left The spectacular movement of the Alter-Real reflects that of its Andalucian ancestors.

Left The Alter-Real's excitable temperament needs careful management, such as that provided by the Portuguese army.

Below By culling undesirable mares and using only the fittest stallions, the Portuguese Ministry of Economy has done much to improve the breed.

The American Saddlebred

Although the American Saddlebred can be used as a general-purpose riding horse (as we shall see, the breed was developed as a working horse), its real role in life is as the 'peacock of the American show-ring'. There, in either the three-gaited or the five-gaited classes, it performs its dramatic paces to the rapturous applause of spectators.

The extravagant paces are encouraged by the use of heavy shoes and the extreme length of the feet.

The canter is of such collection that forward movement is minimal, giving rise to the saying that a Saddlebred can 'canter all day in the shade of an apple tree'. The slower of the 'artificial' paces, the slow-gait, is also highly collected, with a momentary hesitation at its zenith, while the rack is an incredibly fast pace in which horses have been timed at 2 minutes 19 seconds over the mile.

In spite of its highly produced, highly strung and fiery appearance, the Saddlebred is acknowledged as having one of the most amiable temperaments, and is easily ridden and handled by children.

Great attention is paid to the appearance of Saddlebreds; three-gaited horses are shown hogged (roached), and five-gaited with full, flowing mane and tail. A feature of the breed which some people find unattractive is the artificially high tail carriage. This is achieved by nicking the muscles of the dock (illegal in Britain) and setting it in position with a brace or set worn when the horse is stabled.

The Saddlebred is a true product of the way of life in the nineteenth-century southern states of America, specifically in Kentucky, where the breed was formerly known as the Kentucky Saddler. The landed gentry required comfortable, well-mannered horses on which to ride round their huge plantations. The inspection of these estates often lasted from dawn to dusk, so durability was as important as comfort, and then social fashion demanded that the horse should also be elegant.

To achieve these requirements, the Kentucky breeders used Canadian and Narragansett pacers, crossed with Morgans, Thoroughbreds and Arabs, and it is interesting that the blood of the Thoroughbred Messenger, so important in the development of the Standardbred, played a major role in producing the Saddlebred.

The Canadian pacer Tom Hall was one of the great sires, although the stallion Denmark was chosen as the official foundation sire.

Saddlebreds are, of course, exceedingly popular in the United States, and a number have now been exported to Australia.

Height 15hh–16hh approximately.

Colour Black, bay brown, chestnut and occasionally grey. White markings on face and legs are common.

Head Small, well set-on, carried high. Clear-cut features, well-chiselled, straight profile. Large nostrils, fine muzzle. Prominent, full, bright eyes. Small ears, set close.

Neck Long, supple, well crested. Throat-latch clean.

Shoulders Very long, sloping, muscular. High withers extending well into the back.

Body Medium wide, deep chest. Well-sprung, long ribs. Short, straight, broad back. Deep, long, full flanks; broad, round, smooth hips. High-set, well-carried tail.

Limbs (Forelegs) Long, broad muscular forearm; deep, wide knees; short, broad, flat cannons with tendons well veined. Pasterns very long, with 45° slope. Wide heels. (Hindlegs) Thighs full and muscular; muscular broad gaskins. Feet slightly less round than forefeet.

Action The three-gaited horse performs the slow, flat-footed walk, which is a springy four-beat pace; the trot, which is a spectacular, high-actioned two-beat pace; and the canter, which is a slow, rhythmical, highly collected pace. The five-gaited Saddlebred also performs the highly collected 'slow-gait', which is slightly faster than the walk. The 'rack' is a high-speed version of the slow-gait.

Far left The prominent, bright eyes, close-set ears and fine muzzle of the Saddlebred.

Above Plantation owners wanted elegant, comfortable mounts on which to inspect their estates.

Left The high action of the Saddlebred's trot.

The American Standardbred

Below Racing Standardbreds must cover a mile within 2 minutes 20 seconds.

Right Pacers wear a special harness to stop them trotting diagonally.

The American Standardbred was developed solely as a trotting animal, and has been recognized as a breed for less than 100 years. As Thoroughbreds are bred for racing and transmit this ability to their progeny, so Standardbreds are bred for trotting and pacing and 'like begets like' in a quite remarkable way, with trotters and pacers each producing their own kind of movement in their offspring. More robustly built than the thoroughbred, they have outstanding endurance.

Although Standardbreds have an inborn tendency to trot and pace, the excessive speeds at which they perform their respective gaits (up to 30 mph or 48 km/h) do have to be produced by schooling, to overcome the great urge to break into a gallop. This is done by the use of special shoes with weighted toe-clips attached to the toe of the forefeet to encourage the horse to extend his stride. Pacers are discouraged from breaking into a diagonal trot by wearing a special harness which makes it difficult for the animal to move in any way *but* at a pace.

It is generally accepted that pacers are slightly faster than trotters and, before being permitted to race, the horses must be able to cover one mile (1.6 km) in a maximum time of 2 minutes 20 seconds. This figure is known as the 'standard', from which the breed derived its name. Present-day Standardbreds cover the mile in considerably less than this, the fastest taking about 1 minute 52 seconds.

The history of the Standardbred is particularly interesting in that it traces back in direct male line to a single Thoroughbred—Messenger—whose own pedigree includes crosses to all three Thoroughbred foundation sires, but particularly to the Darley and the Godolphin Arabians. Messenger himself goes back to the Darley Arabian through Flying Childers, and he arrived in America from Britain in 1788. He stood at stud for more than twenty seasons, and although he himself never took part in the trotting races then so popular (under saddle, then in harness) in America, he sired many animals who made their names as trotters.

One of his descendants, Hambleton, a big, strong stallion out of a mare with Norfolk trotters *and* Messenger in her ancestry, was foaled in 1849, and he and his progeny so dominated trotting races that almost all other lines died out, and virtually all present-day trotters trace back to him.

The only other breeds that have had any influence at all on Standardbreds are the Morgans and the Clays (animals descended from Barbs). The Standardbreds themselves, however, have influenced harness racers in Russia, Australia, New Zealand, Italy, Holland, Norway, Sweden, West Germany, Denmark and Hungary.

Height 14hh–16hh.
Colour Any solid colour, but mostly brown, bay black or chestnut.
General conformation The Standardbred has the general appearance of a rugged, rather stocky, short-legged Thoroughbred.
Head Moderately refined, intelligent, well shaped. Neat ears and a kind eye.
Neck Elegant but strong.
Shoulders Well sloped and muscular.
Body Of great depth, but tending to be long. High croup, powerful quarters.
Limbs Very strong, short, with plenty of bone.
Action Standardbreds either trot, with the near-fore and off-hindlegs going forward together in the usual diagonal movement, or pace, which is a lateral movement with the near-fore and near-hindlegs going together. The trotter has long, low strides; the pacer has a typical swaying motion.

The Andalucian

The breed description does little to convey the quality and great presence of this attractive breed of little horse. Developed in Spain, almost certainly by the crossing of native mares with Barb stallions that arrived with the Moorish invaders, it became the most popular riding horse in Europe. Its active, brilliant paces made it the outstanding mount for high school riding, and its flowing mane and tail, its symmetry and its elegant good looks ensured it was the first choice for European monarchs when having equestrian portraits painted.

Selective breeding was in the hands of Carthusian monks from about the fifteenth century, and the story of their battle to ensure the breed's purity, and in some instances its actual survival, is one of romance and drama. In time, they created three herds of pure Oriental blood, and so great was their determination to maintain these lines that they defied a Royal Edict directing that foreign, central European blood be introduced into the Spanish studs. Later, they successfully hid enough of their beloved horses from Napoleon's armies to ensure the breed's survival.

To this day, it is claimed that no Arab or any other foreign blood played a part in the breed's development, and the lack of a dished profile tends to support the claim—at least as far as Arab blood is concerned.

There are comparatively few Andalucians left now, and most of these are in Spain, but their former popularity in Europe has ensured that they have had an influence on the equine world that approaches that of the Arab.

They were used in the development of Holsteins, Nonius, Württembergs, Kladrubers, Hannoverians, Alter-Real, the Lusitano and probably even the Connemara pony. The Andalucians were of notable importance in the development of the Lipizzaner, which was founded from Spanish mares and stallions which were taken to Lipizza near Trieste in 1580.

In Spain, the agility and spectacular paces of the Andalucians suit them to the demands of the

Right The Andalucian's straight, rather than dished, profile supports the claim that Arab blood was not used in the breed's development.

Below left The breed's flamboyant paces makes it a natural choice for *haute école* and when crossed with the Thoroughbred a good all-round competition horse is produced.

Below right The agility of the Andalucian makes it an ideal breed to use in Spain's national sport.

Rejoneadores or mounted bullfighters and, when crossed with Thoroughbreds, the extra speed so acquired makes them ideal for the riders testing fighting bulls on the breeding farms.

Apart from the small numbers of Andalucians remaining in Spain, and the recent import of a few individuals into Britain, the breed is not readily available to the average rider in Europe or America.

In Australia, however, the Andalucian, and more especially its first cross with Thoroughbreds, appears to be gaining popularity. Studs exist in five states, and pure- and part-breds are competing successfully in dressage, show jumping and cross-country events. The pure-breds are being used to give demonstrations of *haute école*.

Height 15hh–15.2hh.
Colour Predominantly grey or bay, occasionally black or roan.
Head Handsome, with a straight profile. Broad forehead and large, kindly eye.
Neck Reasonably long and deep, but elegant. Well-formed crest in stallions.
Shoulders Long, sloping and with well-defined withers.

Body Short, strong, with well-sprung ribs and a broad chest.
Hindquarters Very broad, strong and rounded. The tail is set rather low.
Limbs Medium length, clean, elegant and strong.
Action Showy, rhythmical walk. High-stepping trot. Smooth, rocking canter.

The Appaloosa

The most outstanding features of Appaloosas are, of course, their coat patterns, but it must be borne in mind that not every spotted horse is necessarily a registrable Appaloosa. In addition to possessing one of the recognized coat patterns, an Appaloosa *must* have the mottled skin which is clearly seen around the nostrils, lips and genitalia, and it must also have the distinctive white sclera in the eye. Other features which *may* be present are the vertically striped hooves and a fine-haired, rather sparse mane and tail.

Five principal coat patterns are recognized in the breed: *Leopard spotted*—spots of any colour on a light or white background (spots of different colour, such as black and chestnut may be present on the same animal). *Blanket-spotted*—a white rump or back on which there may be spots of any colour. *Snowflake*—white spots on any colour except grey. *Marble*—mottled all over the body. *Frost*—white spots on a dark background. Some Appaloosa foals are born a solid colour (but with the parti-coloured skin and white sclera) and may develop one of the colour patterns later in life, and some are born with one pattern, which changes by the time the animal is one or two years old.

Appaloosas are now bred in many countries, but of course they are best known in America. Here they are extremely popular, especially for Western riding, which requires strong quarters and a quick turn of speed, and for parades, for which their handsome appearance is ideal.

There is now a strong Appaloosa presence in Australia—again being used by the many Western-style riders in that country. In general, the Appaloosas of America and Australia tend to be more of the Quarter Horse type. In 1976 a British Appaloosa Horse Society was formed, with its own rules of judging and its own standards, based on those of the American Society. However, it is interesting that the British Appaloosas are more of the hunter type. They have had some success in dressage, eventing, show jumping, hunting and particularly in endurance riding. There is still a certain amount of prejudice against coloured horses in Britain, but the 'Appies' are showing that they are good performance animals with very good temperaments, so that prejudice may, in time, be overcome.

Although Appaloosas are so closely identified with the American horse world, horses with Appaloosa features pre-date the discovery of the American continent by many thousands of years.

Height 14hh–15.3hh approximately (a minimum of 14.2 hh in Britain).
Head Straight and lean, well set-on. Wide forehead. Nostrils and lips show the typical parti-coloured skin. Ears, pointed and of medium size. The sclera of the eye is white, giving the eye prominence.
Neck Quality, with a clean-cut throat-latch and large windpipe. Mane (and tail) inclined to be sparse.
Shoulders Long, well muscled, sloping. Prominent, well-defined withers.
Body Chest deep but not excessively wide. Short, straight back. Loin short and wide. Underline long with the flank well let-down. Hips are smoothly covered, long, sloping and muscular. Well-rounded quarters.
Limbs Forearm well muscled, long, wide and tapering to a broad knee. Cannons short, wide and flat with wide, smooth and strongly supported fetlocks. Pasterns medium-sloped. Feet are rounded, deep; open and wide at heels, and walls often show the typical vertical black and white stripes. Thighs long, muscular and deep. Gaskins long, wide and muscular, leading to clean, clearly defined wide, straight hocks.

Above A blanket-spotted Appaloosa has a characteristic white rump, on which the spots of the coat clearly show. Appaloosas used for Western-style riding in America and Australia tend to be of the Quarter Horse type.

Above right European Appaloosas generally are of a hunter type.

Above The characteristic vertically striped hoof, which may or may not be present.

Left The Appaloosas of America are extremely popular and have the third largest registry of any breed in the world.

European cave drawings some 20,000 years old show horses with Appaloosa markings, and oriental art dating from centuries BC also show spotted horses.

It is believed that they made their first appearance in North America with the Spaniards in Mexico, but they are particularly associated with the Nez Percé Indians, who 'borrowed' some from the Conquistadors. They introduced a strict breeding policy in their homelands, which was in the vicinity of the Palouse River in Oregon. The horses thus became known as, first 'of Palouse', then 'Apalouse', and finally 'Appaloosa'.

In 1877, in the course of the Indian Wars, the US Army captured the bulk of the Nez Percé as they were fleeing towards the Canadian border, and slaughtered most of their carefully bred horses. A few remained, and in the early 1900s formed the basis of the modern breed, which was virtually saved from extinction by the formation of the Appaloosa Horse Club in 1938. From there, the Appaloosa has gone from strength to strength, and now has the third largest registry of any breed in the world.

The Arab

The Arab holds a position in the horse world that is unchallenged by any other breed. Not only is it generally accepted as the oldest pure breed, it is almost without dispute the most beautiful; it differs anatomically from other horses, and the part it has played in the development of other breeds is unmatched by any other.

Although the origins of the Arab are unknown, there is evidence of its existence in recognizable form in the Arabian peninsula as far back as 5000 BC. There is no doubt that, due to the devoted, almost fanatical determination of the Arabian tribesmen to ensure that no foreign blood should sully its purity, it was established as a breed centuries before any other. In addition, the environment in which it developed instilled in it the remarkable qualities of endurance, soundness, strength, speed and courage, *and* more especially, a prepotency that has ensured that these qualities are passed on from generation to generation.

The Arab is best known for its creation of the Thoroughbred following the importation to England in the seventeenth and eighteenth centuries of the Darley and Godolphin Arabians and the Byerley Turk—the foundation sires of the Thoroughbred breed. It had, however, been imported to mainland Europe far earlier, when the Moors crossed from North Africa to the Iberian Peninsula in the seventh century. In succeeding centuries, the breed spread through the continent, and has played a vital part in the development of almost all the modern European breeds, from the Russian Don to the Austrian Haflinger pony, the Lippizaner, even the Percheron heavy horse, but not, it is said, the Andalucian. It has also influenced almost all the British native pony breeds and, on being transported across the Atlantic with the Spaniards, played an important part in the development of American breeds such as the Morgan. It has, indeed, an extraordinary ability to 'nick' with almost all breeds, and add quality to them.

Pure-bred Arab studs have been established all over the world—particularly in Britain, Europe, America and Australia, and, as might be expected, different types of the breed have evolved. The 'desert' or Egyptian Arab (now, alas, no longer bred by the tribesmen) is noted for its exceptionally refined and graceful head, with great 'dryness' of bone—a feature described as very fine, dense, exquisitely chiselled bone. On the other hand, the desert Arab tends to be a little 'on the leg', a little shallow in the girth and a little short of rein.

Height 16hh max. approximately. Average 14.3hh approximately.
Colour Brown, bay, chestnut, black, grey.
Head Very short, of great refinement. Face has pronounced concavity or 'dish'. Muzzle very tapered with exceptionally fine skin. Nostrils large and markedly elastic. Eyes very large, placed lower in the head than in other breeds; widely spaced. Ears small, fine, mobile and curve inwards. The Arab head is distinctive because of the feature known as the *jibbah*—a shield-shaped bulge extending between the eyes upwards to a point between the ears and down across the top third of the nasal bone.
Neck Of good length. Characterized by the *mitbah*, or angle at which the neck enters the head; it makes a slight angle at the top of the crest, and runs in a gentle curve to the head. Overall an arched curve, allowing the head to move very freely. Mane and forelock of fine, silky hair.
Shoulders Sloping, although not so much as in a Thoroughbred. Well-defined but not very prominent withers.
Back Short and slightly concave. Strength across the loins. Croup long and level. The spine and ribs of Arabs differ from other breeds: the Arab has 17 pairs of ribs, 5 lumbar vertebrae and 16 caudal vertebrae; other breeds have 18 pairs of ribs, 6 lumbar vertebrae and 18 caudal vertebrae.
Body Deep girth and chest. Ribs well rounded.
Hindquarters Generous, high set, silky tail.
Limbs Clean and hard, with well-defined tendons. Bone measurement of the Arab not as a rule substantial; it is said that Arab bone is denser than that of other breeds, and thus stronger. Hindlegs should be strong and well formed.
Action A good, free walk with plenty of movement from the shoulder. The trot should have the wonderful cadenced, floating quality, with the momentary hesitation, which, with the high tail carriage and flowing mane, typifies the uniqueness of Arab movement.

The Spanish Arabs tend to have rather plain heads and straight shoulders, while Polish Arabs, although somewhat plain as youngsters, mature with exceptional shoulders, limbs and quarters.

The modern history of the Arab in Britain revolves around three studs, with the most influential being Crabbet, owned by Wilfred and Lady Anne Blunt. Over the years, a definite Crabbet type emerged, with more substance than most others, and this type has acted as the foundation for the Arabs in America and Australia—although both countries now import from Europe, so the differences are becoming slightly blurred. The other important British studs were Hanstead, founded by Lady Yule, and the Courthouse Stud, founded by Mr H. V. Musgrave Clark.

It might be said that Arabs have done so much to influence the world of horse-breeding that they should not be expected to be high-class perform-

ance animals as well. It is true that they have now been overtaken by the enormous scope demanded in modern equestrian sports such as show jumping and eventing. Their remarkable stamina has, however, made them the endurance riding breed *par excellence*.

While the pure-breds have, perhaps, not been able to respond to the demands of modern sports, the Arab/Thoroughbred cross or Anglo-Arab combines the best features of both breeds, and is often a top-class performance horse.

In spite of the purebred's limitations as a competition animal, it is nonetheless a wonderful riding horse, and for the average rider it has sufficient ability to compete in a variety of events at the lower levels. This, combined with its wonderful temperament, which, in spite of its fiery appearance is delightfully gentle and tractable, ensures that the Arab is in demand throughout the world.

Left The Arab head—arguably the most beautiful equine profile. Among the distinguishing features are the 'dish' in the face, beneath the large, low-set eyes, and the tapered muzzle, small enough to fit into a half-cupped hand.

Above The Arab moves with both rhythmic grace and speed.

Overleaf Arab horses ridden by Fantasia horsemen, in their native habitat in Morocco.

The Cleveland Bay

That Cleveland Bays are not regarded primarily as riding horses is underlined by the fact that they are, in breed classes, always shown in hand and not under saddle. However, their importance in the production of outstanding riding horses when crossed with other breeds (most especially with Thoroughbreds) is such that no account of riding horses would be complete without their inclusion.

Cleveland Bays are noted for their proven ability to transmit to any other breed with which they are crossed the important qualities of substance, outstanding bone, good action, good colour, hardiness and endurance. The pure-breds are good jumpers, and they also pass this on to their stock, producing many fine show jumpers, eventers and especially hunters, when crossed with Thoroughbreds. They are so highly regarded in this context that they have been exported to many other countries, including the United States, Canada, Australia, Pakistan and South Africa.

The pure-bred Cleveland is a very sensible horse, and possesses a strong character which can, if mishandled, be spoiled. The breed is, however, characteristically very bold and honest—traits which are passed on to all stock.

Historically, Clevelands claim to be one of the oldest established British breeds, known at least since medieval times in the Cleveland district of Yorkshire from which they take their name. Originally they were bred as pack horses and for light draught and agricultural work, and came to be known as 'Chapman Horses' because of their use by the chapmen or travelling salesmen of those times.

In Elizabethan times, with the coming of coach travel, their straight and good paces ensured that Clevelands were much in demand as coach horses and, down the centuries, their reputation as carriage horses has grown. To this day, they are used as carriage horses in the Royal Mews, and HRH Prince Philip drives teams of these attractive bay horses with great success in international combined driving events.

In common with other harness and agricultural horses, the coming of mechanization posed serious problems for the survival of the breed, and indeed, as recently as 1962, only four mature stallions remained in Great Britain. Fortunately, the breed underwent a dramatic revival, largely owing to the influence of the great stallion Mulgrave Supreme, who was bought by HM the Queen and, after being broken to saddle and harness, stood at stud until his death.

Right A handsome Cleveland Bay head.

Below When crossed with Thoroughbreds Cleveland Bays produce excellent all-round competition horses.

Height 16hh–16.2hh, but height should not disqualify an otherwise good sort.

Colour Cleveland bays must be bay with black points. Grey hairs in mane and tail do not disqualify; they have long been recognized in certain strains of pure Cleveland blood. White is not admissible beyond a very small star. Legs which are bay or red below the knees and hocks do not disqualify, but are faulty as to colour.

Head Bold, not too small and well set-on. Eyes should be large, well set and kindly in expression. Ears tend to be large and fine.

Neck Long and lean.

Shoulders Sloping, deep and muscular.

Body Should be wide and deep. Back should not be too long, and should be strong with muscular loins.

Hindquarters Should be level, powerful, long and oval. Tail springs well from quarters.

Limbs Forearms, thighs and second thighs should be muscular. Knees and hocks should be large and well closed. There should be 9 inches (23 cm) minimum of good flat bone below the knee, measured at the narrowest point on a tight tape. Pasterns should be strong, sloping and not too long. Legs should be clear of superfluous hair, and as clean and hard as possible.

Feet One of the most important features of the breed; must be of the best and blue in colour. Shallow or narrow feet are undesirable.

Action Action must be true, straight and free. High action is not characteristic of the breed. The Cleveland which moves well and is full of courage will move freely from the shoulder, and will flex his knees and hocks sufficiently. The action required fits the wear-and-tear qualities of the breed.

The Cob

The Cob is a type of horse, not a breed and, although Cobs can be found in many countries, they tend to be typically British, and certainly show Cob classes are a very British institution.

There is no sure method of breeding a Cob, but they are most likely to be the result of crossing a heavyweight hunter (perhaps one with some heavy horse blood) or an Irish Draught horse with a Thoroughbred or near Thoroughbred.

Cobs are sturdy, compact animals, sometimes described as 'heavyweight hunters on short legs', up to a great deal of weight and typically regarded as being ideal mounts for the more elderly, less athletic rider. If this gives the impression that they are dull slugs, it is entirely erroneous. While being calm, steady rides with impeccable manners, they should have a certain gaiety, be active and willing and be able to give their riders an enjoyable, safe day's hunting, or take part in competitive events—perhaps not at the highest level, because their stocky build does impose certain limitations of scope.

Their powerful quarters ensure some jumping ability, while the common sense with which these great characters of the horse world are usually endowed ensures that they will undertake most tasks safely and willingly. The percentage of heavy horse blood in many Cobs allows them to live out more readily than their better-bred relatives.

Cobs are usually shown hogged (roached), and with a full tail; until it was made illegal, they were shown docked.

Cobs of this type should not be confused with the Welsh Cob, which *is* a distinct breed, although Welsh Cobs can and have been used in the production of cob-type riding horses.

Height Up to approximately 16hh, but 15.1hh is the maximum for a show cob in Britain.
Colour Any.
Head Small, of good quality. Any suggestion of coarseness to be avoided.
Neck Strong, arched, elegant.
Shoulders Well sloped, strong and muscular.
Body Short, strong back, with great depth through the girth.

Hindquarters Generous, rounded. Tail set high.
Limbs Very short cannons. Sturdy, with good, not coarse, bone.
Action A good, low action, coming from the shoulder, not the knee. Should be able to gallop well.

Right The cob's comfortable ride and perfect manners give the rider an enjoyable and safe day's hacking or hunting.

The Connemara Pony

Ireland's only native pony, the Connemara is one of the most beautiful of the Mountain and Moorland breeds, and is described by the English Connemara Pony Society as 'unique in the world, combining the strength and hardiness of the mountain pony with the quickness, agility and beauty of the Arab'.

The home of the Connemara is that part of Ireland known as Connaught, bounded in the south by Galway Bay and to the west by the Atlantic. In this wild, beautiful region of mountains, bogs and stony outcrops the hardy pony has developed in a climate which, while harsh in winter, is tempered a little by the Gulf Stream, which enables rushes, herbs, reeds and grasses to appear early in the year.

In spite of the overall poorness of the land, there are vital quantities of phosphates which enrich the herbage and contribute significantly to the strength of the ponies' bones and muscles. For those that live nearer the coast, there is access to the iodine and other minerals in the seaweed, which they eat with relish.

Agility as well as hardiness is a product of the Connemaras' environment; those foaled among the rocky slopes and boggy valleys develop from birth a sure-footedness that stands them in good stead when ridden across any kind of rough, trappy country.

The modern Connemara is ideal for adult and child alike; it is fast, handy, and a great natural

Height 13hh–14.2hh.
Colour Grey, black, brown, dun, with occasional roans and chestnuts. Predominantly grey.
Head Well balanced. Small, neat ears, and large eyes.
Shoulders Long and sloping.
Body Deep and compact.
Limbs Cannons short, clean. 7–8 inches (17.5–20 cm) hard, dense bone.
Action Free, easy and true movement.

jumper, as well as going kindly in harness. Although Irish in origin, the Connemara is very popular indeed in Britain, and in recent years has become firmly established in the United States, Europe and especially Australia, where studs are found from Queensland in the north-east right round to Western Australia.

When crossed with Thoroughbreds, Connemaras produce some exceedingly high-quality, successful performance animals. Show jumping fans will remember with affection little Dundrum, by the Thoroughbred Little Heaven out of a Connemara mare, who, partnered by Tommy Wade, won many international honours, including the King George V Gold Cup. Little Heaven was also the sire of another successful animal, Little Model (also out of a Connemara mare), who represented Britain at dressage in the Rome Olympics. Another great Connemara-cross was Korbous (by a North African Barb), who was ridden into second place at the Harewood Three-Day Event in 1957 by Penny Moreton.

The precise origins of the Connemara are not fully known, but three or four hundred years ago there was almost certainly an infusion of Arab, Barb and Andalucian blood from horses brought in by merchants trading between Galway and the Iberian Peninsula. During the nineteenth century, estate owners in Connemara imported Arabs which bred with the native ponies and stamped on them the

Arab characteristics that are still apparent in the modern breed.

At the end of the nineteenth century, Welsh stallions were introduced, and one of these, the cob Prince Llewellyn, sired a pony called Dynamite out of a native mare, and he in turn sired Cannon Ball, who was registered as No 1 in the Irish Connemara Stud Book.

Sadly, a certain amount of indiscrimate outbreeding also occurred, with appalling results, and it was not until a stud book was started at the suggestion of Professor Ewart that selective breeding restored the pony to its previous quality.

Overleaf The Connemara combines 'the strength and hardiness of the mountain pony with the quickness, agility and beauty of the Arab.'

Left The stamp of the Arab is unmistakable.

Below Connemaras foaled in the wild, harsh conditions of their native environment thrive on poor grazing land, and have the sure-footedness to travel over all kinds of rough country.

The Dales Pony

The Dales pony takes its name from the area of North Yorkshire, Durham and Cumbria where it has been bred for centuries. However, until the end of the nineteenth century it was not recognized as being a separate breed from its near relatives, the Fells, which are bred on the western side of the Pennine range. Since then, the Dales have developed independently, and their height limit has increased to 14.2hh, whereas that of the Fells has remained at 14hh.

Dales are another British native breed which make ideal family ponies, being kindly and tractable enough for a child to ride, while their short, strong backs and general stockiness make them suitable mounts for all but the tallest adults.

Their make and shape does impose some limitations as regards speed, but they are excellent jumpers, can go across country very adequately and are excellent harness ponies. They are also used very successfully in the popular holiday pastime of pony trekking, where their placid temperaments and sure-footedness make them ideal mounts for the many beginner-riders who enjoy this type of holiday.

The Dales' hardy constitution makes them economical to keep and they can live out in the very worst weather that the north of England can offer.

The early history of the Dales and the Fells is identical, with such stallions as Lingcropper and Blooming Heather acknowledged as influential by both breeds. The Dales breeders, however, regard the Welsh Cob Comet as having had the most lasting influence on their breed. A strong, heavy animal, Comet was foaled in 1851, and was taken

Height Not exceeding 14.2hh.
Colour Predominantly black, but bays, browns and, less commonly, greys, are found.
Head Neat and small. Wide between the eyes, which should be bright and docile. Muzzle should be of medium width. Jaw and throat free from coarseness. Ears small, well placed and erect, and the pony should have an alert expression. Mane (and tail) long and thick.
Shoulders Deep, sloping and well laid back.

Body Short, strong back with strong, powerful loins. deep through the girth with great heart room and well-sprung ribs.
Hindquarters Well developed and compact.
Limbs Strong, flat-boned forearms. Short, straight cannons, broad strong and clean-cut hocks. Good clean joints and fine silky feather at the heels.
Feet Broad, very hard, well shaped and blue in colour.
Action Straight and true, with the movement coming from the knees and hocks.

from Wales to Westmorland to compete in trotting races, in which he put up some remarkable performances, once covering 10 miles (16 km) in 33 minutes, carrying 10 stones (63.5 kg). His progeny were recognized as some of the finest in the Dales, with a son, Comet II, being the grandsire of Linnel Comet, said to be the most lovely pony of the heavy Dales type.

Although Dales are now used primarily as pleasure ponies, they come from a long line of working animals. They were used as pack and harness ponies, carrying lead and coal from the mines to the coast, and on farms, where their comparatively small size made them especially useful on the steep hills, where larger animals would have been at a disadvantage. They are still used to a small extent on farms, and some are used for shepherding on the moors.

A small number of Dales are crossed with Thoroughbreds to produce hunters and cobs of great substance and sensible temperament.

Overleaf The typical, alert expression of the Dales pony.

Left The Dales pony uses the power of its strong legs, with movement coming from the knees and hocks, to travel across country.

Below The Dales' placid temperament and sure-footedness makes them ideal for shepherding on the northern hills of Britain.

The Dutch Warm-blood

The very general nature of the above description suggests that the breed has not yet formed any easily recognizable features—and that is, in fact, the case. The Dutch Warm-blood is one of the most recently established of this type of European horse, with a stud book dating from 1958. Nevertheless in that short time the breed has proved capable of producing extremely versatile, high-class animals, for riders of all standards.

The Warm-blood is based on two much older Dutch breeds, the Groningen and the Gelderland. The former was a dual-purpose horse used on farms and as a carriage horse and heavyweight riding animal, whose strong, muscular quarters most attracted the Warm-blood breeders. The Gelderland is a lighter type of animal, and the feature most sought by the Warm-blood breeders is the outstanding forehand.

These two older breeds were crossed with carefully selected Thoroughbreds in the first instance, and later with Warm-blood stallions from France and Germany. The result is an outstanding performance horse of remarkably calm, docile and willing temperament, of rather lighter type than the average *German* Warm-blood, but with more substance than most Thoroughbreds.

The breeding policy of the Dutch Warm-blood Society has been 'to produce a noble and likeable horse with an honest character. It must possess a constitution as strong as steel and have faultless joints so as to guarantee a long life.' The breeders have aimed for a height of approximately 16.2hh, as it was felt that larger horses tend to be less agile

Height Average 16hh, approximately.
Colour Any.
General conformation Good sloping riding shoulders. Pronounced withers. Strong body, muscular quarters. Strong limbs with good bone.
Action Straight, true action, with easy ground-covering strides at all paces.

Above The Dutch Warm-blood is an outstanding performance horse of rather lighter type than the average German Warm-blood.

and, on a strictly practical note, are more expensive to keep. To achieve this, a rigorous programme of selective breeding was undertaken, backed by equally rigorous performance tests for the stallions, and strict conformation appraisals of the mares—the best of which also undergo performance tests.

The stallions, after being chosen initially on bloodlines and conformation, are not used for breeding until they have passed their performance test. Each year the young stallions undergo 100 days of training under identical conditions. The tests at the end include a riding trial, loose jumping, ridden jumping, a harness test and a cross-country test over fixed obstacles and water. The judges award marks for the tests, taking into account character and a report on each animal's stable manners. Those that pass are used as stallions, and when their first crop is born these too are inspected carefully. The mares are selected primarily for conformation and movement, but they are also assessed on their progeny.

The success of the scheme may be judged by the quite remarkable number of horses emerging from the breed that have become international stars, especially in show jumping and dressage. The 1982 Volvo World Cup winner, Calypso, ridden by Melanie Smith for the USA, is a Dutch Warm-blood, as is the incomparable stallion Marius, belonging to Mr John Harding and ridden by British rider Caroline Bradley. In the field of dressage, Dutch Courage, ridden by Jennie Loriston-Clarke to win the Bronze Medal for Britain in the 1978 World Championships, is an outstanding example. The

Dutch whip Tjeerd Velstra drove his team of four Dutch horses to the Bronze Medal at the World Carriage Driving Championships at Windsor in 1980, and in America, Clay Camp also drives Dutch Warm-bloods.

One of the most encouraging features of this comparatively new breed is its willing and placid temperament that makes it suitable for riders who are never likely to reach international heights but want a versatile, pleasant animal to compete at lower levels.

Top The Dutch Warm-blood's versatility makes it equally useful under saddle or in harness.

Above The breed has a willing, placid temperament suitable for riders of varying ambitions.

The Fell Pony

The Fell pony should be constitutionally as hard as iron and show good pony characteristics, with the inimitable appearance of hardiness peculiar to mountain ponies. At the same time it should have a lively and alert appearance and great bone.

Until the end of the nineteenth century, the ponies bred on both sides of the Pennine range in northern England were indistinguishable. Since then, however, those on the east of the Pennines have become a separate breed, known as the Dales, while the ponies bred to the west, and more particularly in Cumbria, have retained their slightly smaller, more compact build, and are known as Fells.

As with the other larger Mountain and Moorland breeds, the Fells are excellent all-purpose ponies for those who require versatility but do not, perhaps, aim to compete at the very highest level. Although the Fell was originally bred for draught and light farm work, it now has a good riding shoulder, and has developed into an excellent

Above The Fell pony's hardy appearance clearly reflects its rugged, inhospitable native mountain habitat.

riding pony. Its strength, sure-footedness (especially over boggy ground) and native good sense make it a comfortable, safe ride for all ages.

Fells can tackle cross-country fences competently, as they are very agile and clever, but they are certainly not built for speed, and lack the scope for top-class junior show jumping. They do, however, perform very well in handy pony classes.

Because of their draught background Fells have really come into their own with the present upsurge of interest in driving. They go exceptionally well in harness, and have often been successful in the inter-breed obstacle driving competition instigated by HRH Prince Philip at the Royal Windsor Horse Show, as well as in the very demanding discipline of combined driving. Fells are also used at trekking centres in many parts of Britain.

The history of the Fells is not without romance. It is thought that when the Romans arrived in northern Britain they found the indigenous ponies too small for hauling the materials needed for building their walls, roads and forts. To infuse more size, they are believed to have imported a number of Friesians to cross with the native ponies, and the resulting stock was said to be very similar to the Fell pony of the present day.

Legend has it that one of the great stallions of the breed, 'Lingcropper', was found, saddled and bridled, 'cropping the ling' on Cross Fell after a border skirmish (perhaps even after the 'Forty Five') in which his rider died. He was taken by a farmer and spent the rest of his life at stud, founding the renowned Lingcropper strain of ponies.

The Industrial Revolution and the subsequent use of mechanized transport almost saw the end of the Fell breed and between the two world wars only five stallions remained. Fortunately, generous support came from, among others, King George V and Beatrix Potter, and a revival was staged. The breed is now firmly re-established.

Above The sturdy Fell was originally bred for farm work but now makes an excellent all-round riding pony.

Height Not exceeding 14hh.
Colour Black, brown, bay and grey, preferably with no white markings, although a star or a little white on the foot is allowed.
Head Small, well chiselled in outline, well set-on; forehead broad, tapering to nose. Nostrils large and expanding. Eyes bright, prominent, mild and intelligent. Ears neatly set, well formed and small. Throat and jaws should be fine, showing no signs of throatiness or coarseness.
Neck Of proportionate length, giving good length of rein, strong and not too heavy. Moderate crest in stallions.

Shoulders Most important, well laid back and sloping. Not too fine at withers, not loaded at the points. Good, long shoulder blade, muscles well developed.
Body Good strong back and outline. Muscular loins. Deep through heart. Round-ribbed from shoulders to flank, short and well coupled.
Hindquarters Square and strong; tail well set-on.
Limbs Forelegs should be straight, well placed, not tied at elbows; great muscularity of arm. Big well-formed knees. Short cannon bone, plenty of good flat bone below the knee—8 inches (20 cm) at

least. Pasterns fairly sloping, not too long. Hindlegs should be very muscular, good thighs and second thighs. Hocks well let-down and clean cut, not sickle- or cow-hocked, plenty of bone below the joint.
Feet Good size, round and well formed, open at heels with characteristic blue horn.
Action Smart and true in walk. Trot, well balanced all round, with good knee and hock action, going well from the shoulder and flexing the hocks, not going too wide or near behind. Should show great pace and endurance, bringing the hindlegs well under the body when moving.

The Gelderland

Height 15.2hh–16hh.
Colour Chestnut, grey, occasionally skewbald.
Head Ram-shaped, rather long ears.
Neck Long and strong.
Body Powerful shoulders, ample chest, wide strong back, deep through heart.
Hindquarters Powerful.
Limbs Fairly short, with good bone. Hard, strong feet.
Action Vigorous and eye-catching.

Gelderlands have not been bred officially since the end of the 1960s, possibly because the Dutch Warmblood, in whose evolution they made a significant contribution, surpassed them as a riding horse, although certainly not as a carriage horse.

The breed had its origins in the nineteenth century in the province of Holland from which it took its name. A variety of stallions, including Norfolk Roadsters and Arabs, were crossed with native mares and produced exceptional carriage horses, which were used by a number of European Royal households.

Later, infusions of Hackney, Oldenburg and Friesian blood were made, producing a very strong, active animal, with the excellent forehand admired by the breeders of the Dutch Warm-bloods, and the elegant action of the carriage horse.

Standing about 15.2hh–16hh, the Gelderlands are usually chestnut or grey, but an occasional skewbald may be found.

Overleaf Infusions of Hackney, Oldenburg and Friesian blood produced the strong, active Gelderland.

Above The Gelderland's excellent forehand.

Above right The stylish action of the carriage horse.

Right Gelderlands ceased to be bred officially in the late 1960s.

The Hackney

Height Hackney horse—14.1hh–15.2hh on average, but may go over 16hh occasionally.
Hackney pony—up to 14hh.

Colour Most usual are dark brown, black, bay and chestnut. Coat is fine and silky.

Head Small, convex. Small muzzle. Large eyes and small ears.

Neck Long and well formed.

Shoulders Powerful with low withers.

Body Compact, with great depth of chest. Tail set and carried high.

Limbs Short legs and strong well-let-down hocks. Well-shaped feet.

Action Brilliance and correctness must always be paramount. Both in action and at rest the horse or pony has highly distinctive characteristics. Shoulder action is free, with a high ground-covering knee action. The foreleg is thrown well forward, not just up and down, with a slight pause of the foot at each stride giving a peculiar grace of movement; an appearance of flying over the ground. The action of the hindlegs is the same to a lesser degree. At rest, the hackney stands firm and foursquare—forelegs straight, hindlegs well back—so that it covers the maximum amount of ground. The head is held high, ears pricked, with a general alertness.

Above right The Hackney pony is virtually a separate breed, not exceeding 14hh.

Right Although primarily a show-ring animal, Hackneys have been driven with success in combined driving events.

Today the Hackney is almost exclusively a harness animal, but when the breed was first developed in the eighteenth century, it was primarily a saddle horse. It was not until roads were improved sufficiently for fast horses to be used in lighter and faster vehicles that they switched roles.

The Hackney breed was developed during the eighteenth and nineteenth centuries in the East Riding of Yorkshire and in East Anglia, using as foundations two related trotting breeds, the Yorkshire Hackney and the Norfolk Roadster. These breeds had a common ancestor in a horse called The Original Shales, by Blaze, a son of Flying Childers, who was by the Darley Arabian. The Original Shales' dam was said to be a Hackney. In those days, the trotting horses developed in the two areas differed from each other, with the Norfolk version being more cob-like and the Yorkshire showing rather more quality. In more recent times the two types have merged, and the best features of both have produced the modern elegant animal.

The distinctive trotting action of the breed is partly inherited, partly taught by the use of, among other things, heavy shoes. If the action is forced too early, it will almost certainly not be sustained in the correct form in the mature animal.

The modern Hackneys are a well-known and popular feature of the British show scene, where their high-stepping, brilliant displays attract great public support. Indeed, show-rings around the world now thrill to Hackney classes, as the breed has been exported in increasing numbers to the United States, Australia, South Africa, Canada and many European countries, especially Holland.

The Hackney *pony* was produced during the late nineteenth century by crossing Hackney horses with native ponies. The result is a real pony which shows the brilliant high-stepping action, possibly to a greater degree than its larger relative.

The Haflinger Pony

The Haflinger is the native pony of Austria, taking its name from Hafling, a village in what is now part of the Italian Tyrol, and which was one of the principal breeding centres. Before the Second World War Haflinger stallions were reared at the Central Stud at Piber in Austria. The breed is believed to have originated from crossing the Alpine Heavy Horse with Arab blood—and it is known that the Arab stallion El Bedavi was used. All today's Haflingers trace back to El Bedavi XXII (a half-bred great grandson) and there has always been quite close in-breeding, resulting in a very definite type with all Haflingers being remarkably similar in appearance.

A mountain breed, Haflingers are very strong and sure-footed, and are used extensively by the farmers and foresters for pack work or light draught where tractors are not practicable. They are not normally worked until they are four years old, and they have a reputation for living to a great age.

Although it is the native pony of Austria, the Haflinger does not, as a rule, display the exceptional hardiness that is associated with the British native breeds. This is said to be because, instead of living out in the severe mountain weather, the ponies are kept in warm conditions, living in stalls or stables under the farmers' houses.

In present-day Austria, the Haflinger has found a new role—that of taking tourists on trekking holidays through the beautiful Tyrolean country-side. They make good riding ponies, well up to weight, and also go very well in harness. A number of the ponies have been exported, especially to Switzerland and Germany, and recently it was announced that the Indian Army mules are to be replaced by Haflingers. A small number of the breed is now in Britain, where the Haflinger Society of Great Britain was founded in 1971. Her Majesty the Queen became one of the first Haflinger owners in Britain, when she was presented with two ponies while on a state visit to Austria.

All ponies that pass the breed society's strict test are branded with the Haflinger mark—a small alpine plant called the edelweiss.

Height 13hh–14.2hh.
Colour Light-, middle-, liver-, and red-chestnut, with a flaxen mane and tail. Red or grey manes and tails are unacceptable. A white star, blaze or stripe is permissible. White on the body or limbs is discouraged.
Head Short with slight dish. Large, dark and lively eyes. Fine nostrils. Small, ears.
Neck Strong, and not too short, suitable for a riding pony.
Body Broad and deep chest; deep girth measuring 67–75 inches (170–190 cm). Well-tensed back. Broad loins with good joints. Muscular croup, not too short. Well-carried tail.
Limbs Clean with hard, hooves. Strong forearms and a good second thigh. Short, strong cannons.

Right The Haflinger's sturdy head and short neck.

Far right Haflingers all look remarkably alike, due to close in-breeding.

The Hannoverian

The Hannoverian is Germany's best known and most successful Warm-blood, with very great talent for show jumping and dressage. There is some variation in size and type, ranging from the big strong heavyweights with massive quarters—from which come their great jumping strength—to smaller, lighter animals. Some of the breed tend to a certain plainness about the head, and there can be narrowness of the feet. There is, however, no doubt about their competitive ability, although they possibly lack the speed to make a really top-class event horse, in spite of their toughness.

The success of the breed in jumping and dressage at the highest level is almost legendary. In show jumping such names as Warwick Rex, individual Gold Medallist at Montreal, and Simona, ridden by the late Hartwig Steenken to the 1974 Men's World Championship, need little introduction, while Gerd Wiltfang's World Championship winner, Roman, is a Westphalian—a Hannoverian bred in Westphalia. The breed is also used in International Carriage Driving competitions. In the 1978 World Dressage championships at Goodwood, two of the three members of the Gold Medal-winning German team were riding Hannoverians—Uwe Schulten-Baumer on Slibovitz, and Harry Boldt on Woyceck.

In common with many European breeds, Hannoverian stallions must pass stiff selection tests and, before being passed for stud duties, they must spend a year in training as riding horses.

The breed is probably descended from the Great Horse of the Middle Ages, but the first record dates back to 1714, when George I of England, who was also Elector of Hanover, introduced British Thoroughbred blood into the German studs. In 1735 George II founded a stud at Celle, for stallions to cross with the mares belonging to local farmers, and initially Holsteins with Neapolitan and Andalucian blood were used. Holsteins were dominant for about thirty years, but then much more Thoroughbred blood was used to produce a lighter, dual-purpose animal capable of working on the farms, but also useful in harness and under saddle.

The Napoleonic Wars, which caused such havoc in the horse population in Europe, were no less devastating to the stud at Celle, where only about thirty of the 100 stallions survived to return to stud when it was re-established at the beginning of the nineteenth century. More Thoroughbred blood was introduced, but this, in due course, had the effect of producing horses that were lighter than desired, so no further infusions were made.

Height Stallions 15.3hh–16.2hh. Mares 15.3hh–16.1hh.
Colour Any solid colour.
Head Medium-sized, clean-cut and expressive. Large lively eye and good free cheek bones.
Neck Long, fine and well placed. Lightly formed at the poll.
Shoulders Large and sloping. Pronounced withers.
Body Strong and deep.
Hindquarters Muscular, with well set-on tail.
Limbs Well muscled. Large, pronounced joints and well-formed, hard hooves.
Action Elastic and energetic, with ground-covering strides coming from the shoulders and hocks, and no high knee action.

Above The Celle stud now has about 200 stallions.

Right The breed's superb ground-covering strides come from the shoulder and great engagement of the hocks.

Overleaf Hannoverians have been bred mainly for equestrian sports since the Second World War.

The Highland Pony

Left The Highland is a remarkably versatile riding pony, suitable for Pony and Riding Club activities.

Right Highlands appear in various shades of dun, as well as grey, brown, black and occasionally bay and liver chestnut.

The Highlands are the largest and the strongest of the British Mountain and Moorland ponies. As their name suggests, they are natives of the Highlands of Scotland, and are also bred in many of the Western Isles. Until recently, the breed was divided into the Western Isles ponies, which were of lighter build and standing up to about 14hh, and the heavier mainland type. The types still exist, but are no longer recognized as such by the Highland Pony Society.

Highland ponies are noted for their very generous manes and tails of fine silky hair, which give them protection from the icy winds and rain of their native heaths, but one of their most interesting features is the wide range of colours to be found. In addition to the grey (which is now the predominant colour) there are, as has been mentioned, a number of shades of dun. The original colour is believed to be yellow dun and, when examined closely, the individual hairs of a pony of this colour are seen to be black at the roots, gradually lightening to a golden shade at the tips, giving an impression of a brilliant gold coat, made more dramatic by the dark ears and a dark face, mane and tail.

Highlands have always been versatile, serving as pack animals, pulling the plough, and being used to take the crofters (small farmers) to market and to carry the peat for the croft fire across rough, boggy ground. They are perhaps best known for their association with deer-stalking, in which their calm sure-footedness makes them ideal for carrying sportsmen up the rugged Scottish hills, and their great strength enables them to carry the shot stags, often weighing up to 16 stones (100 kg), down those same hills.

In recent years, the ponies have become increasingly popular for riding, and in this, too, show a remarkable versatility. Although not the fastest of mounts, they go across country sensibly and safely, are good jumpers, and also go very well in harness. They are handy for their size, and their willing temperaments make them ideal for Pony Club and Riding Club activities. Their calmness is well suited to the needs of the elderly, the nervous and the disabled.

As family ponies they have one overriding advantage—it is often difficult to find keep that is *poor* enough to prevent them from becoming grossly overweight, and so they must be about the most economical animals for their size in terms of food!

In the course of their development, the Highlands have been subjected to infusions of many other bloods, in particular that of Arabian horses, but also that of Clydesdales. Norwegian ponies and even, in one instance, an American trotting horse.

The most important stud in the breed's history is undoubtedly that founded by the Dukes of Atholl. At this stud stood the most influential stallion, the great Herd Laddie. A fine grey pony, he stood at the stud from the age of six, and such was his prepotency that the presence of so many greys in the breed today is attributed almost entirely to him. Herd Laddie was the grandsire of probably the best known Highland pony, Jock, who belonged to King George V, and was often used by him on shooting expeditions.

Height Not exceeding 14.2hh.

Colour Various shades of dun—mouse, yellow, golden, grey cream, fox. Also grey, brown, black and occasionally bay, and liver-chestnut with silver mane and tail. Most ponies carry the characteristic dorsal eel stripe, and many have zebra markings on the inside of the forelegs. Apart from a white star, white markings (blazes, socks, etc.) are disliked and discouraged. Ponies presented for Appendix registration with white markings will no longer be accepted.

Head Well carried; face broad between alert and kindly eyes; short between eyes and muzzle. Muzzle not pinched, nostrils wide.

Neck Strong, not short. Good arched top-line. Throat clean and not fleshy. Mane and tail should be long, silky and flowing, not coarse.

Shoulders Well laid-back with pronounced withers.

Body Compact. Back with slight natural curve. Chest deep, ribs well sprung and carried well back.

Hindquarters Powerful with strong, well-developed thigh and second thigh. Tail set fairly high and carried gaily.

Limbs Flat, hard bone. Forearm strong, knee broad, short cannon, pasterns oblique, not too short. Well-shaped, hard, dark hooves. Forearm placed well under weight of the body. Hocks clean, flat and closely set. Feather silky and not over-heavy, ending in a prominent tuft at the fetlock

The Holstein

The Holstein is one of Germany's oldest breeds of Warm-blood, with its ancestry going back to the Great Horses which carried knights in armour into battle in the Middle Ages. Since then, it has been cleverly adapted to meet the varying uses demanded of it down the centuries.

Originally bred in the Schleswig-Holstein region by putting Oriental and Andalucian stallions to the native mares, as the need for war horses declined and that for harness horses increased, British Yorkshire Coach horses and Cleveland Bay stallions were used in the nineteenth century to produce a fine, high-stepping carriage animal.

More recently, the demand for riding horses increased, and the introduction of Thoroughbred blood produced a handsome, sensible, bold horse with enormous scope, of rather heavier type than the Hannoverian, and having all the best characteristics of a quality hunter. The success of this Thoroughbred infusion can be judged by the quite remarkable versatility shown by the breed at the highest level. In the 1976 Olympics, Holsteiners won an individual gold and individual bronze medal, and team silver medals in all three equestrian disciplines.

Meteor, one of the great show jumpers of the

Top The Holstein's attractive, expressive head.

Left The Holstein brand (in close-up above) dates back to the Great Horse.

Right The breed was recognized in the United States as early as 1892.

post-war years (ridden by Fritz Thiedeman), was a Holsteiner, and more recently the brilliant dressage horse Granat, ridden by the Swiss Christine Stuckleburger, has been in a class of his own for a number of years. In the horse trials discipline, Madrigal, ridden for Germany by Karl Schultz, won the individual Bronze Medal at Montreal in 1976 and the Wylye Horse Trials in Britain in 1979.

Most Holsteiners are perhaps a little larger than the average rider would find strictly necessary, but nevertheless, their versatility, and their kind, sensible temperaments make them excellent all-purpose horses.

The breed has been exported to many countries, and the United States government recognized the stud book as early as 1892.

Height 16hh–17hh. Average height for three-year-old filly eligible for registration, 16hh approximately. Young premium stallions at 2½ years, 16.1hh–16.2hh.

Colour All colours permissible. Bay with black points and brown are the most usual. Grey is quite common, chestnut less so.

Head Expressive, well set-on and in proportion to the size of the horse. Big, bright eyes. Lower jaw should not be too heavy.

Neck Long, muscular, slightly arched.

Shoulders Long and sloping.

Body Wide, deep chest. Back strong. Loins muscular.

Hindquarters Strong and muscular. Tail carried well though not too high set. Strong, muscular thighs, stifles and gaskins.

Limbs Good flat knees and big clean hocks. Short, strong cannon bones. Medium-length pasterns sloped in correlation to the shoulders. Forelegs should be well apart and elbows must have freedom of movement. A stallion should have between 8 and 9½ inches (20 and 24 cm) of bone.

Feet Adequate size, open at the heel, dense and smooth.

Action The walk should be long, straight, free and elastic. The trot must be balanced, free-going and cover a lot of ground. The canter must be easy, well balanced, smooth and straight.

The Hunter

In general terms, a hunter is any horse (or pony) that will carry a rider safely and comfortably to hounds, and it is thus a type, and not a breed as such. In Britain, the term 'hunter' evokes a picture of an animal standing something over 15.2hh, of good riding conformation and some quality, with sufficient bone, substance and jumping ability to stand up to the rigours of a day's hunting.

Different types of hunter are found, depending to a large extent on the nature of the country over which they are to be hunted. For example, in the English Shires, where there is good galloping grassland, a Thoroughbred or near-Thoroughbred is ideal, whereas in a country where there is a great deal of plough, a strong, short-legged animal is more suitable.

The late R. S. Summerhayes described the ideal hunter in the following terms:

'He must, of course, be absolutely sound and stand on the best of legs; his body must be generous and sufficiently ample to allow heart, lungs and so on to perform their duties under conditions of great exertion, and further, he should give his rider as long a rein as possible. His head must be of the right size and his neck obviously of the correct length to assist with the many acts of balancing which he must perform during a hunt. Almost of more importance, if it is possible, the high-class hunter must be courageous and bold, tireless and, as it is said, always able to find "an extra leg" if in trouble. He must not chance his fences, but must stand back and boldly attack each one as he meets it; such being the case, it is obvious that if the right temperament is there the ideal horse to hunt is the Thoroughbred, or as near as may be to one that is Thoroughbred?'

That the Thoroughbred is the foundation of the best British hunters has long been recognized by the Hunters' Improvement and National Light Horse Breeding Society (the HIS). They appreciate that the enormous stud fees commanded by Thoroughbred studs connected with the racing industry are beyond the means of the average hunter breeder, and they therefore introduced a system of stallion premiums. By means of this scheme, the services of Thoroughbred stallions, carefully selected for conformation, action and suitability for producing quality horses of size and substance, are made available to members' mares at reasonable cost. The success of this scheme may be judged from the fact that many progeny of Premium stallions have won top honours not only in hunter show classes but also every kind of equestrian sport. Such well-known horses as HRH Princess Anne's Goodwill, Anneli Drummond-Hay's Merely-

a-Monarch, and Diana Mason's Special Edition are all by Premium stallions.

There are a variety of show classes for hunters in Britain, the principal ones being for lightweights—capable of carrying up to 13 stones (83 kg), the middleweights—capable of carrying up to 14 stones (89 kg), and the heavyweights—capable of carrying over 14 stones (89 kg). The classes are judged on conformation, action and ride but, unlike similar classes in America, they are not judged on jumping ability. This, in Britain, is restricted to working hunter classes. In all classes, however, the ability to gallop on *and* come back are considered of extreme importance.

Although, as has been described, many horses are bred specifically to be hunters or top-class competition horses of hunter-type, a great many riding horses can be classified as being of hunter-type, although of no special breeding. They are often very versatile, taking part in a variety of events in addition to hunting. Many Riding Club members own hunter-type horses with the necessary bone and substance to take part in cross-country events, hunter trials and show jumping, as well as being handy enough to compete in gymkhana-type classes, and having enough stamina to compete in long-distance rides.

Above The 'hunter' is a type of horse or pony of good riding conformation, with the strength of build, intelligence and agility to withstand a day's hunting.

Right The British hunter is a versatile animal, taking part in a variety of activities, such as cross-country events and hunter trials, in addition to hunting.

The Irish Draught

The name 'Irish Draught' is slightly misleading, for these horses can be (and are) ridden, and show considerable aptitude for jumping. Nevertheless they were originally used on farms in Ireland for light draught work as well as for riding. The breed's most important use now, however, is for crossing with Thoroughbreds to produce top-class up-to-weight hunters and competition horses that have been exported from Ireland to many countries all over the world.

The origin of the Irish Draught is unknown, but it is thought that it may have evolved from Connemara ponies which, removed from their own bleaker habitat, were bred up on the lusher keep in other parts of Ireland. It is also possible that the Irish Draught resulted from the use of imported Thoroughbred stallions on some of the native mares, which themselves had been infused with Spanish or Arabian blood.

Prior to 1850 the breed was described as 'a long, low build of animal, rarely exceeding 15.3 or 16

Height 15hh–17hh.
Colour Bay, grey, brown, chestnut.
Head Small, intelligent and of hunter type.
Body Deep chested with sloping shoulders. Oval rib cage. Short smooth coat. High-set tail.
Limbs Clean legs with good, flat bone, and feet that are not over-large. There should not be any feathering.
Action Free and straight.

Above The Irish Draught may have evolved from Ireland's native pony, the Connemara.

Left The Irish Draught is somewhat misleadingly named, as it shows a great aptitude for jumping.

MRS. BOYLAN
29. ARD NA CRUSHA (1)

hands high with short, strong, clean legs, plenty of bone and substance, short back, strong loins and quarters, but the latter inclined to be goose-rumped, slightly upright shoulders, strong neck and smallish head'. They had good strong and level action, without being extravagant, could trot, canter and gallop well, and were excellent jumpers.

However, the Irish Famine of 1847 seriously depleted the breed, and those remaining were rather coarsened by infusions of Clydesdale and Shire blood. Later, the Department of Agriculture introduced subsidies to approved Irish Draught and hunter-type stallions, and this eventually re-

sulted in the production of the good, active, useful animal of today.

In 1917 a stud book was started, and in 1976 the Irish Draught Horse Society was formed to promote and preserve the breed. A number of stallions were exported to England, where they had long been regarded as a *type* rather than a breed. In 1979 a British Irish Draught Horse Society was formed with the same aim as its counterpart in Ireland. The British Society has introduced a strict inspection scheme for stallions and mares, and a registration system which includes a section for animals of Irish Draught type but which have no papers.

Above When crossed with Thoroughbreds, Irish Draughts make top-class competition horses.

The Lipizzaner

As Thoroughbreds are to racing, so Lipizzaners are to the art of *haute école* or High-School riding, that form of equitation made famous world-wide by the Spanish Riding School of Vienna. The Lipizzaner is a beautiful, spectacular horse, nearly always grey and possessed of great grace, dignity and a unique athleticism that makes it particularly suitable for the brilliant 'airs above the ground'—the *croupade*, *courbette* and the *capriole*.

The very attributes, however, that make it so suitable for *haute école* have possibly limited it for more general use, as it rather lacks the ability to extend freely, which is demanded in other equestrian activities. Nonetheless, it is a fine carriage horse and the skills of 1980 World Carriage Driving Champion Gyorgy Bardos were undoubtedly enhanced by the agility and manoeuvrability of his brilliant team of Lipizzaners. In addition, their comparatively small size give them a distinct advantage over larger breeds when negotiating the very difficult and tight obstacles in the marathon stage.

Although the Lipizzaners have been associated with the Spanish Riding School in Vienna, the name of the school gives the clue to their origin. The foundation stock, said to be the descendants of the Spanish horses used on ceremonial occasions by Roman emperors, were of Andalucian origin. Nine stallions and twenty-four mares were imported into Austria from Spain by the Archduke Charles II to found a stud at Lipizza, near Trieste, which, although now in Yugoslavia, was then part of the Austrian Empire.

The stud at Lipizza, including, tragically, all the documents, was all but destroyed during the

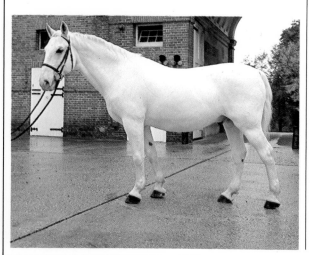

Height 14.3hh–16hh, averaging 15hh–15.2hh.
Colour Mainly grey, occasionally bay. Foals born black or dark brown.
Head Well shaped, often a ram-like profile. Large eyes. Narrow nostrils.
Neck Well set-on with abundant mane.
Body Strong, compact, muscular body, deep through the girth.
Hindquarters Very powerful. Tail set high, with thick, silky hair.
Limbs Short cannons, dense bone. Flat joints, well-formed feet.

Above Lipizzaners are nearly always grey. The head is well shaped and the eyes are large and expressive.

Far left The stamp of the Andalucian is evident.

Left A brand on the left croup—one of four with which each foal is marked at six months.

Napoleonic Wars. Fortunately, the 300 horses had been moved out of harm's way. They returned, only to be forced to flee again in 1805, this time to Slavonia. Although they returned once more in 1815, tremendous harm had been done to the breed in that decade, and serious degeneration had taken place. Careful breeding restored the former quality, and Franz Joseph I ordered that the 'Lippies' were to remain at Lipizza where the breeding conditions were ideal, and this they did until the outbreak of World War I.

Once more on their travels, the mature stallions and four-year-old mares were taken to a stud near Vienna, while the remainder went to an old-established stud at Kladrub in Bohemia. When Lipizza eventually became part of Italy, the horses were divided between Italy and Austria, with those belonging to Austria going to their present home at Piber in West Styria. The remainder went again to Lipizza.

The story of the Lipizzaners' last move was told to the world in a film, 'The Flight of the White Stallions', which traced their flight from Hostau in Czechoslovakia, where those from both Lipizza and Piber had been taken, to the safety of Bavaria. From there they were eventually returned to Piber.

Over the centuries, six great stallions emerged, and it is on these that the breed has been based, and all modern-day horses bear the prefixes of their lines. The first great stallion was Pluto, born of pure Spanish origin in 1765 at the Royal Danish Court stud at Fredericksborg. Next came the black Neapolitan Conversano, also of Spanish descent; in 1779 the dun Favory was born at the stud in Kladrub; Neopolitano, another Neapolitan, was foaled in 1770 in Poesina, in Italy. The white Arabian Siglavy was born in 1810, while the last, Maestoso, was born in 1819 by a Neapolitan sire out of a Spanish mare.

The pedigree of each foal can be read from four brands with which it is marked at the age of six months. A 'P' for Piber with the Austrian Imperial Crown over it is marked on the left croup; an 'L' for Lipizzaner is on the left cheek; on the left side of the back, two letters are branded, indicating the sire's line and the dam's sire—'C' for Conversano, 'F' for Favory, 'M' for Maestoso, 'S' for Siglavy, 'P' for Pluto and 'N' for Neopolitano. The foal's registration number is branded on the right under the saddle.

Above A Lipizzaner of the Spanish Riding School, which has made the breed famous, performing *capriole*, one of the 'airs above the ground'.

The Lusitano

Height 15hh–16hh.
Colour Predominantly grey, but any solid colour is permitted.
Head Small, straight profile. Small ears. Bold eyes.
Neck Tends to be short and rather thick.
Shoulders Long, sloping. Withers not particularly well defined.
Body Short-coupled, strong, broad-chested. Tail set rather low.
Limbs Strong, clean and of medium length.
Action Tends to be showy, but less so than the Andalucian.

Far left Making use of their grace and agility, Lusitanos are now bred for dressage and, in their native Portugal, for bull-fighting.

Above The Lusitano looks very like the Andalucian, on which it was based.

Left Lusitanos were at one time used as replacement horses by the Portuguese army.

The Portuguese Lusitano is not dissimilar in appearance to the Spanish Andalucian, although it has rather less natural presence. It is an ancient breed, based on the Andalucian, and was used for light farm work as well as under saddle.

Lusitanos are good riding horses, and their grace and agility make them particularly suitable for dressage and, in their native country, for the Portuguese style of bull-fighting. This is conducted entirely on horseback, with the horses being very highly schooled to swerve instantly when given the aids as the charging bull approaches. If a horse is injured in the Portuguese bull-ring, it is considered a terrible disgrace.

They are now being bred in Suffolk, England, by Lord and Lady Loch, who train their horses for Grand Prix dressage.

The Morgan Horse

Height 14.2hh–15.2hh approximately.
Colour Brown, bay, black, occasionally chestnut.
Head Medium size, clean-cut and tapering slightly from jaw to muzzle. Profile can be straight or slightly dished, never Roman-nosed. Wide, clean-cut lower jaw. Medium fine muzzle with small, firm lips and large nostrils. Ears should be small, fine-pointed, set wide apart and carried alertly. Eyes should be full, bright and clear.
Neck Medium length, well crested. Clean-cut at the throat-latch. Smoothly joined to shoulder and deep at the point of the shoulder. Mane and forelock good and full.
Shoulders Good length and slope, blending into smooth, well-defined but not too high withers.
Body Chest of good depth and width. Back short, broad and well muscled. Loins should be wide and muscular and closely coupled. Barrel large and rather round. Well-muscled quarters.
Limbs Forelegs should be short, squarely set, well apart. Short muscular forearms, and short, wide, flat cannons. Fetlock joints should not be round but rather wide. Pasterns clean and strong, of medium length, the slope to correlate with slope of shoulder. Hooves of medium size, nearly round. Hocks wide, deep and clean.
Action The walk should be flat-footed, elastic and rapid with a long straight, free stride. The trot must be square, free-going, collected and balanced. The canter must be smooth, easy, collected and straight on both leads.

The Morgan is unique in being the only breed to be descended from, and named after, a single horse. That horse, originally named Figure, came to be known as 'Justin Morgan's horse' after his owner, a poor singing teacher in eighteenth-century New England.

The little horse, who stood only 14hh and weighed about 800 lb (360 kg), was acquired as a two-year-old by Justin Morgan in about 1795, and proved to be an animal of quite exceptional strength and versatility.

In spite of his diminutive size, he was entered in log-pulling contests (allegedly never losing) and pulled loads that horses of 1,200 lb (545 kg) or more could not move. He worked at ploughing, in harness, and at the heaviest kind of land clearance, tearing rocks and tree stumps from the rugged landscape.

Justin Morgan's horse also took part in many races, usually run over a quarter of a mile (440 m), and it is said that he was never beaten, either under saddle or in harness.

In addition, he was widely used as a stallion, and his prepotency was such that he has passed on, through successive generations, his remarkable strength, speed, endurance and wonderfully gentle temperament.

There is considerable uncertainty about the horse's breeding. Some authorities suggest he was by a Dutch-bred stallion, but in general he is believed to be by a horse called True Briton. This horse was said to be a Thoroughbred, but an interesting theory put forward by the late Miss Pauline Taylor, an authority on Welsh Cobs, suggested that True Briton might have been a Cob—particularly as the name is a very common one in the annals of that breed. The statue of Justin Morgan's horse at the Morgan Horse farm in Vermont shows an animal with many cob features, and the present Morgans are, in many respects, very like Welsh Cobs. Possibly one of the most noticeable features is the cob-like high head carriage, known in the Morgan breed as 'upheadedness'.

Today's Morgans, just like their great progenitor, are remarkably versatile, going exceptionally well under saddle and in harness, and are said to rival animals with Arab blood for endurance in long-distance riding. Their elegance and presence make them ideal animals for *concours d'elegance* classes in private driving. They are great jumpers, and Arete, ridden to a show jumping Gold Medal by the

late General Humberto Mariles in the 1948 London Olympics, was a Morgan.

Over the years, two types of Morgan have evolved—the Park horse and the Pleasure horse. Although the conformation is basically the same, the Park horses (so named because they were originally ridden in the parks of the southern American states, much as elegant hacks were ridden in London's Hyde Park) develop an exceptionally high-stepping trot, moving with great animation and what is called 'attitude'—the ability to

derive great enjoyment from displaying themselves. The full, high trot is not wholly developed until the horse is about seven years old, and it is said not to be possible to breed them to order. The offspring of a great Park mare and stallion will not necessarily inherit their attitude, and it is not a characteristic of the Morgan temperament that can be produced by training.

Morgans are, of course, an American breed, but there are now a number in Britain, making names for themselves in various fields.

Left The elegant action of the Morgan can be put to use in harness, as well as under saddle.

Above The American Morgan owes its strength, speed, beautiful movement and gentle nature to its single ancestor, 'Justin Morgan's horse' of Vermont.

The Mustang

The mustang, most familiar as the wild 'bronco' and the cowboy's horse in American Westerns, was the original 'cow-pony'.

Of Arabian and Barb ancestry, the breed was introduced into South and then North America by the invading conquistadors in the 1500s. The Red Indians bred and selected Mustangs as their Indian horses, and by the 1800s large herds of strays roamed wild or semi-wild on the plains. The original Mustang was a small, very tough, wild-tempered horse, seldom exceeding 14.2hh. Infusions of Thoroughbred, Morgan, Quarter Horse and other blood produced its successor, the range horse, although the true Mustang is still used as a 'cow-pony' and makes a good light saddle horse.

Right A result of cross-breeding, the range horse has succeeded the Mustang.

Below Mustangs are the feral and semi-feral horses of West and South America.

The New Forest Pony

In spite of its considerable height range, the New Forest pony is generally regarded as a large Mountain and Moorland breed for show purposes, and most breeders endeavour to produce strong animals with plenty of bone, at least over 13hh.

The Forester is versatile, and has the scope to tackle most equestrian activities successfully. Over the years, New Forest ponies have competed successfully in dressage, long-distance riding, show jumping, eventing and in harness, and although their size precludes them from most senior events, a number have done well in senior Riding Club competitions.

Foresters are, in common with the other native breeds, extremely hardy, economical to keep, and possess the calm, kind temperament so desirable in a family pony.

The breed's native habitat is the New Forest in Hampshire, where, at present, some 2,000 or more roam, in the heather, bogs and poor pastures, in a semi-wild state. They belong to the New Forest Commoners, whose properties bear ancient rights of grazing. Every autumn, the spectacular pony 'drifts' or round-ups are held. The New Forest Agisters (who are, under the New Forest Verderers, responsible for the day-to-day welfare of the ponies) organize the drifts. On the appointed day the Agister for the particular area, accompanied by the Commoners owning ponies in that area, sets out to round up the animals. All are mounted on fast ponies, and with a great deal of shouting and cracking of whips, they herd the ponies into pounds or railed enclosures. There they are wormed and checked, and the Agister cuts their tails in a special pattern to show that the 'marking fee' (the fee payable for grazing) has been paid. Foals to be sold are removed, and the mares and

Height 14.2hh (upper limit). There is no lower limit, but ponies are seldom under 12hh.

Colour Any except piebald, skewbald and blue-eyed cream. Bay and brown predominate. White markings on head and legs are permitted.

General characteristics A New Forester should have a pony head, well set-on, long sloping shoulders, strong quarters, plenty of bone, good depth of body, straight limbs, and good hard, round feet.

Action This should be free, active and straight, but not exaggerated.

other foals turned back into the Forest again for another year.

Every spring and autumn, sales of ponies are held at Beaulieu Road sale yards near Lyndhurst, and one thing can be guaranteed—any New Forest pony taken from the Forest and sold will be traffic proof, as it has, literally from birth, been accustomed to the traffic on the roads of the Forest.

Of all the British native breeds, the New Forest is the least distinctive in type. This is because, until as recently as the 1930s, an enormous variety of different breeds were turned out in the Forest, some indiscriminately, others in a conscious effort to improve the bone and substance of the Forest ponies, which had degenerated seriously during the course of several centuries.

At various times, Arabs, Thoroughbreds and Hackneys were used to 'improve' the local stock, and examples of almost all the other native breeds, from Highland to Welsh, were turned out and have left their mark on the breed. To this day, it is possible to see ponies with distinct Arab features, and even the occasional 'mealy nosed' signs of the Exmoor. It is said that almost all the dun ponies now running in the Forest trace back to one Highland pony.

Many New Forest ponies have been exported, with the majority going to Western Europe, but there are also sizeable populations in Canada, the United States and, more recently, in Australia and New Zealand. In Australia, a Forester has been accepted for registration in the Australian Stock Horse Stud book—further proof, if any were needed, of the breed's versatility.

Overleaf A mixture of almost all British native breeds, more than 2,000 New Forest ponies live in a semi-wild state in the county of Hampshire, England.

Above Its docile nature and ability to be easily trained makes the Forester an ideal family pony. Larger ponies, while narrow enough for children, are quite capable of carrying adults. The smaller ponies, however, often show more quality.

The Norwegian Fjord Pony

The Norwegian Fjord pony is an attractive yet primitive-looking pony that has inhabited Norway since prehistoric times. Originating in the west of the country, it was well known to the Vikings, who used it for the unpleasant sport of horse fighting.

The Fjords are still used extensively in Norway as farm ponies on the smallholdings in the mountains, where they can work in places inaccessible to tractors. They are particularly sure-footed and hardy, which make them especially suitable for the mountainous regions. The Fjords perform well as pack ponies, and their friendly, docile characters have made them popular in a number of European countries, both under saddle and in harness. There are now a few Norwegian Fjord ponies in Britain.

Height 13hh–14hh.
Colour Dun, with black eel stripe down centre of back; occasionally zebra stripes on the legs. Very occasionally bays and browns are found. Mane and tail are composed of black hairs in the centre and silver hairs on the outside. The mane stands stiffly upright and is cut in a characteristic curve.

Head Pony head. Large, well-spaced eyes and a broad flat forehead. Ears small and widely spaced.
Neck Tends to be short and thick set.
Body Short, compact and heavily muscled. Withers rather short and round. Tail set rather low.
Limbs Small amount of feather on heels. Short cannons.

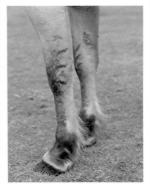

Top Fjords make perfect work ponies in the mountains

Above Zebra stripes are often seen on the legs.

Left The primitive-looking breed has inhabited Norway since prehistoric times.

The Oldenburg

Height 16.2hh, approximately, upwards.
Colour Brown, black and bay, some greys.
Head Rather large, and may have a Roman nose.
Neck Moderately long, thick and extremely strong.
Shoulders Well muscled, sloping.
Body Deep, well muscled, broad chested and inclined to be flat-ribbed.
Hindquarters Powerful, with high-set tail.
Limbs Short, with plenty of bone, and large joints.
Action Fairly high knee action.

Far left This warm-blood stands over 16.2 hh.

Left Oldenburg's rather large, plain head.

Below Crossing with Thoroughbreds has produced an all-purpose horse.

Overleaf Oldenburgs are again being bred for harness.

The massive Oldenburg, often standing well over 17hh, is Germany's largest and heaviest warm-blood. Based on Friesian blood, the breed was founded some 300 years ago by Count von Oldenburg, who infused lighter blood, particularly from Spain and Italy, to satisfy the demand for coach horses. The quality of the resulting crosses was improved further in the second half of the eighteenth century by the infusion of Spanish, Barb and English half-bred blood to give a strong harness horse, which, due to the Friesian blood, matured early.

With the ending of the great demand for harness horses, breeders were forced to change the type yet again. Two stallions in particular, the Thorough-bred Lupus and the Norman Condor, produced a useful riding horse. More recently, further infusions of Thoroughbred blood, together with some Hannoverian and Trakehner, have established the breed as a fine all-purpose horse, but perhaps lacking a little in hardiness and endurance.

In a sense, however, the wheel has come full circle, for the Oldenburg, originally bred as a coach horse, is now making its name again in harness. The British Crown Equerry, Sir John Miller, drives a team of Oldenburgs from the Royal Mews, as do the Germans Bernd Duen and Ernst Fauth in International Combined Driving competitions.

The Palomino

There can be few more spectacular sights in the equine world than that of the glinting golden coat of the Palomino, set off by the striking white of mane and tail. No wonder that the colour has been valued and nurtured for centuries—and the fact that it is not possible to breed it with any certainty has presented breeders with a challenge many of them cannot resist.

In the United States, the Palomino is recognized as a colour breed and certain standards, listed below, have been laid down. In addition, there are restrictions on size that virtually eliminate the inclusion of pony breeds as registered Palominos.

In Britain and in many other countries, Palominos are recognized as a colour type, and *not* as a breed. In Britain, any Palomino-coloured riding animal may be registered with the British Palomino Society, including the native pony breeds, some of which, such as the Welsh Mountain, the New Forest and the Shetland, permit such colouring.

The Palomino Horse Association was formed in America in 1936, with the aim of producing a distinct breed and to record blood lines and provide certificates of registration. However, genetic studies suggest that it is most unlikely that Palominos will ever breed completely 'true' and, although this has done nothing to diminish the popularity of the colour, it has perhaps discouraged owners in America from registering their horses.

Because of a genetic phenomenon known as 'incomplete dominance' it is not possible to produce Palominos with any certainty, even when like is mated with like. However, there are at least four types of mating that will produce a known percentage of the colour. When two Palominos are mated, two Palomino foals will be produced to one chestnut and one albino; the mating of a Palomino with a chestnut produces, on average, one Palomino to one chestnut; a Palomino with an albino produces one Palomino to one albino. The mating of chestnut to Palomino *does* always produce Palominos, but the colour is frequently very poor, and is not favoured.

Because, world-wide, there are so many Palominos from so many different breeds, they have made a name for themselves in almost every aspect of riding. There are Quarter Horse Palominos that excel in Western riding, part-Arab Palominos that compete in British show classes, and others of no particular breed that do well under saddle and in harness.

(The following breed description is the official standard of the Palomino Horse Association of America.)

Height 14.1hh min.–16hh max.

Colour Near that of an untarnished gold coin; may be lighter or darker but must be natural. Must have a full white mane and a natural white tail.

General conformation Good head. Wide between the eyes, which are dark or hazel, and both the same colour. Small alert ears. Should stand on well-set legs, with good withers, a short straight back and a good natural tail carriage.

Markings White on the face permitted but not to exceed blaze, strip or star. White on legs is acceptable to the knees or hocks.

Skin pigmentation May be either of dark or golden colour. The Palomino Horse Association does not discriminate against either skin colour and endeavours to discourage controversies over this basic, natural difference. It is the firm belief of the Palomino Horse Association that discrimination as to skin colour can come only through personal difference of opinion rather than scientific reasoning.

Above The Palomino colour—near that of a gold coin—is very ancient. Although it is non-existent in present-day Arabs, it is thought to have originated in some of the earliest examples of the Arab breed.

Left The Palomino, in America a registrable colour breed, is a popular choice for Western riding enthusiasts. Because of a genetic phenomenon it is impossible to breed Palominos with any certainty, a challenge many breeders find irresistible.

Overleaf In Britain, the Palomino is considered a colour type rather than a breed and interest in the type has only relatively recently been fostered by the British Palomino Society. Any horse or pony with the golden colouring, white mane and tail, and dark but not blue eyes is eligible for registration and classes are now held for Palomino-coloured animals at horse shows.

The Quarter Horse

A sleepy little critter that can unwind like lightnin'' is a much-quoted and apt description of the American Quarter Horse. It sums up exactly the outstanding features of these remarkable horses, with their docile, friendly character, and their astonishing ability to take off from a standstill into a blistering gallop that can leave even a Thoroughbred panting in the rear over the quarter-mile distance from which they take their name.

In spite of being associated in many people's minds with the American West, the Quarter Horse had its origins in the Eastern states and the early European colonization of that area. The Spaniards brought Barbs (and possibly Arabs), while the later settlers from England brought in Galloways from Scotland and, later still, Thoroughbreds.

The American Indians stole horses from the Spaniards and, in due course, the English obtained some of them and crossed their Thoroughbreds or near-Thoroughbreds with these 'Chickasaw' ponies to produce a tough, strong animal which developed into what is known as the Quarter Horse.

The British settlers did not only bring their horses with them; they also brought their love of racing. There were no formal race-tracks, so they made their own wherever they could: on roads, or over a necessarily short distance cleared in woodland—a distance that in time was standardized to about a quarter of a mile (440 m). Over this distance the compact, muscular little horses with their phenomenal sprinting ability excelled.

In the harsh pioneering world, however, horses were not used for racing alone. They had to be jacks-of-all-trades, existing under difficult conditions. They worked on the farms, hauled timber, were used in harness and, most important of all, they worked the cattle. They were ideally suited to the latter task. Their stocky, compact build enabled them to twist and turn with astonishing agility, and their sprinting ability allowed them to race and head off recalcitrant cattle with a burst of speed that no other breed has been able to match. Over the years they developed a remarkable instinct or 'cow savvy', and to watch a good stock horse 'working' cattle is one of the gripping spectacles of the equine world. The horse, head down, eyes fixed on the animal it is working, anticipating every move, turning, swaying, spinning, without aid from the rider, needs to be seen to be appreciated. There is nothing quite like it.

The Quarter Horse is also a versatile pleasure horse, and there are few equestrian sports it cannot tackle with success. There are Quarter Horses competing in show jumping, endurance riding, dressage, polo, and in show classes; they are seen in the hunting field and, most spectacular of all, in rodeo and all forms of Western riding, including barrel-racing and cattle-cutting.

Their popularity can be judged by the fact that the Quarter Horse Registry is the largest in the world, with well over 1,500,000 recorded.

Quarter Horses have been exported to a number of countries, and have been extremely successful in Australia, where Western riding and rodeos are now big business. There are Quarter Horses in Britain, but they have never really made a great impact.

Height 14hh–16hh at maturity.

Colour The American Quarter Horse Association will not accept for registration any animal having one or more spots of such size, kind and in such location as to indicate Pinto, Appaloosa or Albino breeding.

Head Relatively short and wide with a small muzzle and a shallow, firm mouth. Upper and lower teeth meet when biting. Well-developed jaws imply great strength. Nostrils full and sensitive. Ears medium length, alert, and set wide apart. Eyes set wide, reflecting intelligence and placid disposition.

Neck The head joins the neck at an angle of approximately 45°. Throat-latch trim without too much thickness or depth, and there is width between the lower edge of the jawbone to enable him to work with his head down without restricting breathing. Neck of sufficient length blends into sloping shoulders. A high-arched neck or a heavy crest is considered undesirable.

Shoulders Long, set at an angle of about 45° to give a long stride. Smooth, but relatively heavily muscled. Slope of the shoulder blends into the medium high and well-defined withers. Withers and croup approximately the same height.

Body Deep and broad-chested, giving great heart, girth and wide-set forelegs. Muscling on the inside of the forelegs gives the appearance of a well-defined inverted V. Back short, close-coupled, full and powerful across the loin. Barrel deep with well-sprung ribs. Underline longer than the back and does not cut high into the flank.

Hindquarters Broad, deep and heavy when viewed from side or rear. Muscles full through thigh, stifle, gaskin and down to the hock. Croup long and slopes gently from hip to tail. Loins blend into croup. Hip muscling long, extending down into stifle, which ties in well to the gaskin. Gaskin muscle extends down into the hock joint, both inside and outside. The stifle is deep and when viewed from the rear is the widest part of the animal, extending out below the hip and above the gaskin.

Limbs Hocks broad, flat, clean, strong, low-set and free of excess tissue. Muscling ties well into the hock joint. No play or give in the joint except directly forward. Cannons short, with hock and knee joints low to the ground. Both front and rear show a perpendicular position and appear quite broad when viewed from side. Tendons appear sharply separated from the bones and from each other. Fetlock is well formed and strong to withstand shock and strain. Medium-length pastern denotes strength; have a slightly forward slope of about 45°. Viewed from front or rear the legs, cannons and pasterns are straight. Hoof is oblong and its size balances with overall size of the animal; tough textured with deep, wide, open heel, and has same slope as pastern.

Top The Quarter Horse is ideally suited to working cattle. To see its superb manoeuvrability and acquired 'cow savvy' in action is unforgettable.

Above The upper and lower teeth should meet when the horse bites.

The Selle Français

Selle Français (French saddle horse) is the name which, since 1958, has been used for the varied collection of riding half-breds found in the regions of France such as Angevin, Charolais and Vendée. The largest and most successful breed contributing to the Selle Français was the Anglo-Norman, and the Selle Français stud-book is a continuation of the Anglo-Norman.

The Anglo-Norman itself was descended from the heavy Norman draught horse, which has been known for a thousand years, and which was, in the seventeenth century, crossed with German, Arab and Barb blood. Later, in the eighteenth and nineteenth centuries, Norfolk trotter and Thoroughbred blood was used, and the Anglo-Norman resulted. Two types developed: the French trotter and a general-purpose horse that was used firstly for carriage work, then in the cavalry and finally as a more general riding horse.

Today some 33 per cent of horses called Selle Français are sired by Thoroughbreds, 20 per cent by Anglo-Arabs, 45 per cent by Selle Français and 2 per cent by French trotters. Thoroughbreds which are crossed with French trotters, Arabs or Anglo-Arabs interbred with French trotters, and Thoroughbreds crossed with Anglo-Arabs (provided there is less than 25 per cent Arab blood) may also be registered as Selle Français.

The Selle Français has been developed primarily as a competition horse, particularly as a show jumper, of which Gilles Bertram de Balanda's Galoubet Malesan and Patrick Caron's Gai Sarda are prime examples. The breed is also used for racing, cross-country racing and eventing. Most are good hunter-types with a spirited but kindly nature.

The weight classification mentioned below is based on the horse's ability to carry a medium or heavyweight rider. The mediumweights are the most numerous and the more highly bred.

Height Classified, as three-year-olds, according to weight-carrying capacity. *Mediumweight*: Small, 15.3hh and under; medium, over 15.3hh but not over 16.1hh; large, over 16.1hh. *Heavyweight*: small, under 16hh, large 16hh and over.
Colour Commonly chestnut, but all colours permitted.
Head Fine, with wide-set eyes and long ears.

Neck Long and elegant.
Shoulders A good, sloping riding-type.
Body Strong, deep-chested, well-sprung ribs.
Hindquarters Powerful.
Limbs Long, with plenty of bone.
Action Lively, supple and a good stride.

Right The breed's fine head and elegant neck.

The Swedish Warm-blood

The Swedish Warm-blood (sometimes also called the Swedish half-bred) is an excellent all-round riding horse of considerable quality. It appears to be well suited to competition riding, and its equable temperament makes it suitable for riders of all standards—up to and including the Olympics. The breed has a special aptitude for dressage, producing such well-known international horses as Piaff, Junker, San Fernando, Werder and Widin in Europe, and Fellow Traveller, ridden by Linda Zang for the United States.

Swedish Warm-bloods have also reached the top in three-day eventing, with Iluster winning the individual Gold Medal in the 1956 Olympics and Sarejevo the Bronze in the Munich Olympics. At a slightly less exalted level, but underlining the breed's versatility, a Warm-blood won the working hunter championship in Montreal twice in the 1960s.

Selective breeding, beginning at the Royal Stud at Flyinge over 300 years ago, has been the key to the Swedish Warm-bloods' success. Originally, Spanish, Oriental and Friesian stallions were imported with the aim of producing horses for the royal stables and for the cavalry. More recently the all-round ability of the breed has been enhanced by the use of Arab, Hannoverian, Trakehner and Thoroughbred stallions—the latter being represented by animals of the quality of Nasrullah's half-brother, Darbhanga.

The stud at Flyinge is still the principal base for the Warm-bloods, and the breeding remains highly selective—both for animals at the stud and for those in private ownership.

A selection system known as the 'Preimerings' is in operation. Before being used for breeding purposes, all mares and stallions must be passed by inspectors, who travel round the country every spring choosing suitable breeding stock. Top-quality animals are classed as 'A', the next 'B', and others, less outstanding, as 'approved'. Reclassification is always possible if a mare of stallion produces outstanding progeny, or they themselves prove good competition horses. Similarly, they can be down-graded, and a stallion can be removed from the approved list if he consistently sires poor stock. All stallions are not only approved for conformation, soundness and so on, but must also pass a performance test.

The breeding stock at Flyinge is augmented each autumn when approximately fifteen colts of the current year's crop, bred by private breeders, are

bought with the intention of standing them at the State Stud. They are inspected again as yearlings, and those considered unsuitable are gelded and sold at auction. The remainder are inspected again as 2½-year-olds, and the best are broken and schooled on for their performance tests. The remainder are discarded.

In this way the quality and performance of the breed are maintained and enhanced.

Height 16.2hh max. approximately.
Colour Any solid colour.
General conformation Good riding characteristics, easy paces and an amenable temperament. Plenty of depth through the girth. Short, strong legs. Up to weight.

Left Breeding of Swedish Warm-bloods remains highly selective to preserve quality and performance.

Above Like other European warm-bloods, the Swedish is particularly suited to the discipline of dressage.

The Tennessee Walking Horse

The Tennessee Walking Horse is yet another interesting American breed that was developed as a specialized working horse and is now renowned as a specialized show-ring performer. As with the Saddlebred, it was developed to carry a rider on inspection tours around Southern plantations, and while it cannot compete with the Saddlebred's *five* gaits, it has two unique ones (the flat walk and the running walk), and a canter that differs in details from that pace in other riding horses.

The official description of the gaits is as follows:

The flat walk and famed running walk are both a basic, loose four-cornered lick, a 1-2-3-4 beat with each of the horse's feet hitting the ground separately at regular intervals (the near-fore, then off-hind, off-front, near-hind). As he moves, his head will nod in rhythm with the regular rise and fall of his hooves, overstriding with his hind foot the track left by his front foot—near-hind over near-fore, and off-hind over off-fore. In general, he should travel in a straight, direct motion, never winging, crossing or swinging. The flat walk should be loose, bold and square, with plenty of shoulder motion. The running walk should be executed also with loose ease of movement, pulling with the fore feet, pushing and driving with the hind. There should be a noticeable difference in the rate of speed between the flat walk and the running walk, but a good running walk should never allow proper form to be sacrificed for excessive speed. The rocking chair canter is a high, rolling motion with distinct head motion, chin tucked and a smoother, collected movement.

The description is, of course, accurate, but it gives little idea of the astonishing 'feel' of the paces of a Walker. It has been described as 'a fluid, smooth, gliding gait' during which the rider is able to sit almost motionless. This is because of the use of the horse's legs from the elbow, rather than the full shoulder so much sought in most other riding horses. The gaits are natural to the breed and can be seen in foals.

At the running walk, the horse normally travels at between 6–8 mph (9.5–13 km/h), but in the show-ring speeds considerably in excess of this are reached, with the very real possibility of overreach injuries. As in the Saddlebreds, the Walker's tail muscles are nicked.

In addition to the two gaits, the Tennessee Walker does, of course, perform the ordinary walk, trot and canter and, as a versatile pleasure horse, has gained widespread popularity in the United States, not least because of its exceptionally kind, docile temperament. The combination of gentleness and comfort have made the Walker the ideal

Height 15hh–16hh.
Colour Black and solid colours the most popular, but any is permissible, and there may be skewbalds, piebalds, etc.
Head Tends to be quite large and inclined to plainness. Eyes distinctive, with eyelids that are often wrinkled and sloping.
Neck Strong and arched; rather shorter than the Saddlebred, which it closely resembles.
Shoulders Sloping. Withers not particularly sharply defined.
Body Powerful. Broad chest and strong quarters. Moderately long back.
Limbs Clean and hard.

Above The Tennessee Walking Horse has a rather large, plain head, and is less elegantly built than its counterpart, the American Standardbred. Its docile temperament and relaxed ride are, however, gaining it increased popularity, especially for beginners.

horse for the beginner, the nervous or the elderly, and the breed is also used for trail and endurance riding, as well as going well in harness.

It is evident, from what has been said, that the Tennessee Walker and the Saddlebred have much in common. Both breeds were used for inspecting the large Southern plantations—and the Tennessee Walker was also known as The Plantation Walking Horse or, because the overseers often inspected the plantations row by row, as 'turn row' horses.

Selective breeding using Morgans, Thoroughbreds, Arabs and Standardbreds produced the desired qualities of both the Saddlers and the

Walkers and, until the beginning of this century, little, if any, distinction was made between the two. By 1910, however, the Tennessee Walker was beginning to be recognized as a breed in its own right, and in 1935 the Tennessee Walking Horse Breeders' Association established a register. In the course of so doing, research showed that most horses were descended from a single stallion known as Black Allan, foaled in 1886, the progeny of a trotting stallion, Allendorf, and a Morgan mare, Maggie Marshal. Black Allan, more formally named Allen F–1, was thus declared the foundation sire of the Tennessee Walking Horse breed.

Above American 'Plantation Walking Horses' were bred as a comfortable way for farmers and planters to inspect their Southern estates. Because the horse moves its legs from the elbow rather than the shoulder, its walking paces are so smooth that the rider sits almost motionless.

The Thoroughbred

Height 14.2hh–17hh approximately. Average approximately 16hh.
Colour Any solid colour, with bays, browns and chestnuts predominating. White markings allowed.
Head Very refined and intelligent. Big, bold eyes. Medium-length ears.
Neck Elegant, arched neck.
Shoulders Markedly sloped. Well-defined, prominent withers.
Body Deep through, giving plenty of room for heart and lungs. Short back, well-sprung ribs. Strong, muscular quarters. High croup, tail set high.
Limbs Hard legs with minimum of 8 inches (20 cm) of bone (except in the very smallest). Well-let-down hocks.
Action Long, low, raking stride, covering a great deal of ground.

Far left The Thoroughbred is the racing machine *par excellence*. As racing varies in its demands in different countries, so those countries have produced their own stamp of horse.

Left The Thoroughbred head is outstandingly refined and intelligent.

Since the seventeenth and eighteenth centuries, the Thoroughbred has been developed in Britain as the supreme 'racing machine', which is now capable of covering a mile at something approaching 40 mph (64 km/h), and shorter distances at correspondingly faster speeds. The name 'Thoroughbred' is the literal translation of the Arabic *Kehilan*, meaning purebred, and it is widely known that the English Thoroughbred traces back to three Arab stallions imported in the late seventeenth and early eighteenth centuries. These three, the Byerley Turk, the Darley Arabian and the Godolphin Arabian, were crossed with native and other mares and produced the forerunners of the modern Thoroughbred. Although it is frequently overlooked, other stallions such as the Leedes Arabian and the Lister Turk also played a part in the early days of the breed, but the three 'founding fathers' take precedence as it is possible to trace direct tail-male lines back to them.

Within fifty years of the arrival of the Byerley Turk in Britain in 1689, Thoroughbreds were being taken across the Atlantic by the early settlers—the first recorded being Bulle Rock, a son of the Darley Arabian, who arrived in 1730. From these, the vast American racing industry of today has evolved. Since those days, Thoroughbreds have been exported worldwide—indeed, wherever there is racing there are Thoroughbreds.

Although the elegant long-legged Thoroughbreds, with their great presence and quality, are easily recognizable, there are, as in other breeds, different types. These have been developed in answer to the demand for sprinters or for middle- or long-distance horses. The sprinters, who need to generate enormous acceleration over short distances, are usually short-coupled, with shorter musculature (as in those explosive sprinters, the Quarter Horses), while the stayers are usually

longer and rangier, with longer muscles and bones.

In Europe, the accent has been on middle- and long-distance horses, while in America the pressure for quick results has resulted in the development of speed and early precocity, although the trend is now more towards the longer-distance animal. Australia, too, produced youngsters that were great sprinters, so that many of their middle- and long-distance races were won by the marvellous types produced in New Zealand. Recently, however, Australian Thoroughbreds have earned an enviable reputation for stamina and toughness.

All this may seem not to have a great deal of relevance to the horse ridden by the average rider, but that is not so. The majority of Thoroughbreds are still bred for the race-track, but it is those which are not suitable for racing that so often find their way into what may be called, for want of a better description, 'private ownership'.

In America, the show jumping and eventing riders have clearly taken advantage of the current production of horses with great stamina. For a decade or more, the Americans have been noted for the wonderful class horses with bone, substance and quality on which they have competed at international level. On the whole, these horses have been quite long-bodied, but very elegant, especially about the head and neck, and have exhibited extreme toughness. They have also been beautifully schooled.

The Australians, when they do ride clean-bred horses, appear to favour small, tough animals that can jump like stags.

In Britain, Thoroughbreds of the more substantial, galloping, hunter type have been extremely successful in show jumping and eventing at the highest level, but, with one or two exceptions, their temperaments have rather let them down under the constraints of top-class dressage. As hunters, in galloping country such as the English Shires, they are without peers.

For the general rider, however, the Thoroughbred—beautiful, high-couraged and versatile though it is—is not necessarily the most suitable. Apart from the fact that it is not very economical to keep, it is so often a highly strung, flighty animal, requiring very careful handling.

Perhaps more suitable is the Thoroughbred with, for instance, some Cleveland Bay, Irish Draught or native pony blood. The outcross instils rather more hardiness and, in general, a more equable temperament (while retaining the elegant paces and the jumping ability) which is better suited to the requirements of the rider who wants a useful all-purpose horse to enjoy.

Having mentioned Thoroughbred crosses, it is appropriate also to mention that the Thoroughbred has exerted enormous influence, second only to the Arab (and perhaps equal to the Andalucian), in the development of other breeds, including the Trakehner, the Selle Français and the Holstein. Any breed that has, in the course of its evolution, required an infusion of quality, has almost always turned to the Thoroughbred.

The Trakehner

The Trakehner (formerly known as the West Prussian) is generally regarded as the most elegant and handsome of the German Warm-bloods. A quality riding horse, with the refined head of a Thoroughbred and wonderful action, it is renowned for its kindly intelligence and for its great stamina and endurance. Perhaps a little light of bone compared with the build, for instance, of the Hannoverian, that bone is nonetheless of fine quality.

As might be expected, the Trakehner has shown great talent for dressage, eventing and show jumping. Show jumping fans whose memories go back to the Stockholm Olympics in 1956 will recall Hans Gunter Winkler's great Trakehner mare Halla. In the first round of the individual championship Winkler had pulled his riding-muscle, and was in such pain during the second round that all he could do was sit there and steer. The great-hearted Halla took her crippled rider round the course clear to win the Gold Medal.

As with so many continental breeds, the Trakehner has had a chequered career, and the story of its survival in what is now West Germany is not without drama.

In the early part of the eighteenth century, Friedrich Wilhelm I of Prussia decided to establish a stud in the north-western part of East Prussia. He set 600 soldiers to drain a vast area of swampy ground—a task that took some six years—and turned the wasteland into well-drained plains which, supplied with lime and phosphorus, produced the fine bone for which the breed is renowned.

He also supplied the foundation stock of animals from other Royal Studs and selected examples of the Schweiken horses that were native to East

Height 16hh–16.2hh.
Colour Any solid colour.
Head Refined. Large intelligent eyes, and a small muzzle.
Neck Elegant and tapering.
Shoulders Well sloped.
Body Medium length, well ribbed-up and strong.
Hindquarters Well rounded.
Limbs Hard, with short cannons and excellent feet.
Action Straight and true with great freedom at all paces.

Prussia. To these he added a number of top-quality Arabs imported from Poland and established the breed that took its name from Trakehnen—the name of the stud itself.

Frederick the Great, son of the stud's founder, also took a great interest in the horses, but after his death in 1786 the stud was taken over by the state, with a subsequent modification of breeding policy to produce cavalry horses.

The stud survived various wars in the area, although with some depletion of stock and, just before World War II, the Germans introduced a very high-class Polish Arab to lighten the stock again. Trakehners were also bred outside the stud, and at the outbreak of war there were some 25,000 mares belonging to 15,000 breeders.

The war was no less disastrous to the Trakehner breed than it was to the rest of Europe. As it drew to its close, the Russians advanced towards the main breeding area, and the owners, desperate to save themselves and their prized horses, loaded their belongings onto wagons, harnessed some of the horses to pull them, rode the stallions and drove the mares towards what is now West Germany.

Of the many thousands of Trakehners, only about 1,200 reached the West, and in the intervening years the Germans have nurtured and bred them on, so that now they are virtually the only national (as distinct from regional) horse in the Republic, and have influenced several of the other Warm-blood breeds. They have been exported to a number of countries, including Britain.

Left The Trakehner is generally accepted as the most elegant and handsome of all German Warm-bloods.

Above The Trakehner action shows great freedom of movement at all paces.

The Welsh Cob, and the Welsh Pony of Cob type

Left The cob's spectacular action, though free, true and forcible, is not always the straightest.

Above A Welsh Section C yearling colt—the Welsh pony of Cob type, which is a smaller version of the Welsh cob, standing up to 13.2hh.

Above A Welsh Section D stallion—a real Welsh cob. The breed is being used with Thoroughbreds to produce excellent

hunters, particularly suited to trappy country, and high-quality performance animals famed for their trotting ability.

Height The Welsh Pony of Cob type (Section C): up to 13.2hh. The Welsh Cob (Section D): minimum 14.2hh; average 14.2hh–15.1hh approximately (no upper limit).

Colour Any solid colour.

Head Full of quality and pony character. A coarse or Roman nose is most objectionable. Eyes bold, prominent and set widely apart. Ears well-set.

Neck Lengthy and well carried. Moderately lean in the case of mares, but inclined to be cresty in the case of mature stallions.

Shoulders Strong, but well laid-back.

Body Back and loins muscular, strong and well coupled. Deep through the heart and well ribbed-up.

Hindquarters Lengthy and strong. Ragged or drooping quarters are objectionable. Tail well set-on.

Limbs Forelegs set square and not tied-in at the elbows. Long, strong forearms. Knees well developed with an abundance of bone beneath them. Pasterns of proportionate slope and length. When in the rough a moderate quantity of silky feather is not objected to, but coarse, wiry hair is a definite objection. Hindlegs must not be too bent and hocks not set behind a line falling from the point of the quarter to the fetlock joint. Second thighs strong and muscular. Hocks large, flat and clean, with points prominent, turning neither inwards nor outwards. Pasterns of proportionate slope and length.

Feet Well shaped. Hooves dense.

Action Free, true and forcible. The knee should be bent and the whole foreleg should be extended straight from the shoulder and as far forward as possible in the trot. Hocks flexed under the body with straight and powerful leverage.

Of the four breeds of Welsh Pony—the Welsh Mountain pony, the Welsh Pony, the Welsh Pony of Cob type and the Welsh Cob—only the last two are really up to enough weight to make family ponies. Both are extremely strong, active animals, powerful jumpers, bold, willing, but with the delightful temperament which characterizes all the British native ponies. The Section C pony has tended to be overshadowed by the largest and strongest of the Welsh breeds, the Section D, the latter being described, with some justification, as the best ride and drive animal in the world.

Both share with the other native breeds the hardiness that enables them to be kept out all the year round, one of the reasons why they make very suitable family ponies.

Both breeds are of ancient origin, with evidence that they existed as far back as medieval times. Their exact ancestry is lost in antiquity, but it seems certain that the Welsh Mountain pony contributed to their development. The Section D Cob has for centuries been a family animal, very much a part of Welsh country life. It was strong enough to work on the farms, ploughing, working in harness and, of course, as a means of transport both in harness and under saddle. Before World War I, the Germans bought large numbers of Section D Cobs, and during the war both the British and the German armies used them.

The great pace of Cobs has always been the trot, and even today, some breeders still regard the canter with suspicion! Until 1918, when stallion licensing was introduced, stallions and breeding stock were chosen by the simple method of holding trotting matches over long distances. It must be said that the Cob, spectacular though it unquestionably is, is not the straightest of movers.

In modern times, there have been four stallions which have influenced the Section D Cobs, and one or more of them appear in the pedigrees of most of the best present-day animals. They are: Trotting Comet, born in 1836, by Flyer; Cymro Llwyd, a dun, by an Arab out of a Welsh trotting mare; Alonzo the Brave, who was foaled in 1866 of Hackney descent; and True Briton, born in 1830 by a Yorkshire Coach Horse out of an Arab mare who was alleged, unlikely though it sounds, to look like a Welsh Cob! The infusions of the various high-stepping breeds such as the Hackney and the Yorkshire Coach Horse have certainly left their mark on the Cobs.

The Section C Cobs have had a slightly chequered history in modern times. After World War II, there were just three stallions in existence, but happily the breed has increased in popularity and, although still far smaller numerically than the Section Ds, it is now in good heart.

Welsh Section Ds are being used with Thoroughbreds to produce excellent hunters and high-quality performance horses, many with a special aptitude for dressage.

Cobs have been exported to various countries, including Australia and New Zealand, but have not yet reached the enormous popularity of their smaller relatives, the Welsh Mountain ponies and the Welsh ponies.

The Württemberg

Height 16hh max. approximately.
Colour Brown, black, bay, chestnut.
General conformation A strong, stocky riding horse, tending almost to cobbiness. Has great depth through the girth and excellent limbs.

Left The modern Württemberg is a good ride and drive animal.

Below The modern breed was greatly influenced by an Anglo-Norman stallion called Faust, who sired many horses with the required hardiness and calm temperament.

Right The Royal Stud at Marbach was founded in the 16th century.

The Württemberg was originally developed as a hardy, amenable all-purpose animal for use on the mountainous holdings of the Württemberg region of Germany. Initially, local mares were put to Arab stallions from the Royal Stud at Marbach, which was founded in the sixteenth century. Later a variety of other blood was used, including East Prussian, Norman, Nonius and Oldenburg. For a short time, heavy horse blood was introduced with the use of Suffolks and Clydesdales.

Although the breed's origins go back nearly 400 years, Württembergs have only been bred systematically for about a century, and a stud book was begun in 1895.

Since the end of World War II, the use of the breed has changed from work horse to sporting horse, and more quality has been introduced by greater use of East Prussian blood (with its background of Thoroughbred and Arab). The most influential stallion of recent years was the East Prussian Trakehner Jumond, who stood at Marbach from 1960.

The present-day Württemburg is a good, sensible, riding-type animal that also goes happily in harness. It is noted for its strength, soundness of legs and feet and excellent temperament. A hardy animal, it is said to be an economical feeder.

The Competitive Spirit

THE AUTHORS OF the chapters in this section are experts who 'do' as well as write. Their aim is to steer you along the road they have taken as well as laying down the guidelines for further progress. The styles of riding described vary from the classical art of dressage to the demanding skills required for show-jumping and eventing. In between it covers side saddle riding, showing, hunting, and western riding plus the increasingly popular sport of driving.

EVENTING

Eventing is the complete test of all-round horsemanship. It combines dressage, cross-country and show jumping, and an event may be held over one, two or three days, with a cross-country course of 1½–4 miles (2.5–6.5 km) and twenty or thirty solid fences varying in difficulty to suit the standard of event and the grade of horse competing. The grades range from Novice or Preliminary (pretraining and training in the United States), through Intermediate to Open or Advanced.

The three-day event is the ultimate test. The cross-country section on the second day is preceded by a steeplechase course of five to ten fences over 1–2 miles (1.5–3 km) and two roads and tracks sections varying from about 6–10 miles (10–16 km) overall. These last complete what is called the speed and endurance phase, the most exciting part of the whole event. The dressage test on the first day and show jumping course on the third day complete the three-day event and provide an equestrian version of the modern pentathlon.

The right type

The very nature of the sport demands that you start with the right type of horse if you want to reach the top. He must be physically tough enough and mentally bold enough to take on the formidable cross-country fences, yet be calm and obedient in the dressage and show jumping arenas. Fifteen hands is the minimum height for adult eventing, and several 'ponies' of this height, such as Australia's Our Solo, Britain's Our Nobby and America's Marcus Aurelias, have proved that size is not everything.

Nevertheless, the ideal horse for beginners to this tough and demanding sport is a five- or six-year-old Thoroughbred or three-quarter-bred of 16–16½hh with hard, flat bone, correct conformation and movement, strong hindquarters and good feet. He should be deep through the girth (a sign of stamina), move well enough to produce a pleasing dressage test when trained, and of course be able to cope with the size of fences at which you are aiming. A calm temperament is best, yet he should

have the fire that produces that little bit extra when the chips are down.

The training programme

Early fitness Having got your horse, the most important factor now becomes his training and getting him fit so that you can both cope with the competition. Further comment on fittening is given in the stable management chapter, but if your horse has ever shown any tendency to tendon problems in the past or has been out at grass for several months prior to starting work, it would be as well to spend a couple of weeks longer at this slow stage. In climates where the ground is always firm, on the other hand, it may not be necessary to do as much hardening-up work because horses adapt to their environment and will normally have harder legs; in this case too much work at the trot could be detrimental because of the concussion.

Schooling on the flat The next stage is best spent hacking out and schooling the horse on the flat to get him really obedient and supple. At least a further fortnight should be allowed here before you are tempted to do any jumping. The event horse must be obedient and alert if he is going to be a safe conveyance across country and this is the moment to ensure that he is learning to do exactly what you want of him. The section on dressage goes into more detail about the type of work that needs to be aimed at for the dressage phase. What the event judges are looking for is a good workmanlike performance: they are not expecting brilliance, though they would undoubtedly be pleased to be confronted with it. They expect to see an accurate, active and supple horse performing the set movements obediently.

Jumping training The ideal way to start is with some grid work to make him more athletic over his fences and teach him to cope with a variety of different strides. He must be able to shorten and lengthen his stride and your grids should be aimed at teaching him to do this. As every horse will have a different length of stride, it is important that the distances between the poles or cavaletti are right

for your horse at this stage. Exercises are suggested in the section on show jumping.

Other training and fitness work Be very careful not to overdo your sessions of jumping practice. Untold harm is done by asking too much too soon. It must be stressed that grid work is quite demanding, so never make the jumps too big nor ask for too much in one session and always end on a good note. Jumping sessions two or three times a week will be ample; other work should include rides out when the horse should be asked to pop over small cross-country fences such as logs and ditches which are safe, suitable and not too big for his stage of training. Flat work should also be continued daily as this is the foundation to all other types of work.

When possible trot and canter up and down hills to teach your horse to balance himself. He must also learn to cope with different types of going so that he is as happy on slightly rough ground as he is on good going.

Cross-country schooling Confidence is the most important factor in cross-country riding so just allow the horse to enjoy himself without over-doing it and don't go fast until he is sufficiently experienced at negotiating all types of fence at a steady pace. It is often a good idea to go cross-country schooling with a more experienced horse who can give you a lead over anything that your horse may be worried about until he really gets the hang of things. It is a shame to worry a young horse unnecessarily if he lacks a bit of confidence, and very often one session with another horse to help out when in difficulties (which are almost bound to arise at some stage) may make all the difference. It is important not to do this too often, however, as the horse comes to rely on a lead.

The idea when schooling for cross-country is to familiarize your horse with as many different types of fence as possible so that when he comes across them in competition he can cope. Your grid work should have taught him to be supple enough to deal with most types of combination fence, but he must also know how to jump on and off banks and up small steps. He must be happy jumping fences at an angle, which is often necessary across country—but never attempt parallels on the angle.

Overleaf Big, upright parallels need bold but accurate riding.

Above Although a cross-country phase should exert the greatest influence over the final result of an event, a careless moment in the show jumping can not only drop a competitor down several places, but can lose a competition.

Right The event horse will often have to negotiate water, either by going through it or over it. Allow the novice horse to splosh around in water whenever the opportunity arises and if there is a small log or drop, so much the better.

Young or inexperienced horses and timid types will need more time to learn how to deal with all these problems so that they gain the confidence necessary. Wait until you are sure your horse has this confidence before setting off to compete.

General care of the horse

Feeding During all the training and schooling it must never be forgotten that the general care of the horse is of the utmost importance. Feeding plays a very important part and the happy medium is often difficult to achieve with the event horse, as he needs enough food to enable him to cope with competitions without stress, yet remain sensible enough to perform a quiet dressage. This is even more important when the horse is in training for a three-day event and is that much fitter and possibly more liable to playfulness.

Shoeing and foot protection Shoeing is of paramount importance. The feet and legs of any horse, and particularly an eventer, are continuously subjected to a lot of wear and tear. Regular shoeing to keep the feet at the correct angle and length will avoid a lot of disappointment later on. If the toes are allowed to get too long it will put more strain on the back tendons. Discuss your horse's movement with the farrier as slight corrective shoeing will usually prevent those maddening knocks or overreaches which can put you out of action.

Few horses move absolutely straight and if you know your horse's weak points it is possible to take appropriate action such as always using protective boots behind if he tends to knock himself there, and overreach or bell boots to prevent bruised heels for jumping and so on. Prevention must always be better than cure.

Fast work When your horse is fit he will need some fast work. Remember to think ahead if you plan a gallop and don't expect him to be clear in his wind if he has just eaten and had a net of hay. Give a little less hay the night before and remove his net a couple of hours before you intend to work him. Galloping a horse is quite an art from the riding point of view, but it is amazing how often people expect their horses to gallop on a full stomach. *You* wouldn't be able to run very fast just after a meal without getting very puffed!

Competitive practice

When you feel your horse is fit, has done enough flat work and is performing confidently over fences, it is a good idea to go in for a few competitions. These need not be events; you could try show jumping on its own or dressage with jumping, just to make sure that you and your horse are going well together. It may be that you can do some clear-round jumping or choose a suitable class that will have the right-sized course for your standard. The first outing of the season is best used as a school unless you are sure all is going well, as you are bound to be a bit rusty after a break, and your horse will need a quiet introduction.

The cross-country can be practised either at hunter trials or at cross-country training outings. Don't go too fast but let your horse find his own rhythm and try to keep him at that speed. If it feels slow don't worry too much—you can make up a lot of time by riding really accurately from one fence to the next; the tearaways waste time by being unable to turn. If you have the fast, overexuberant sort of horse, don't fight him too much. It is better to allow him to go on a bit so that he settles sooner: which he will, as long as you don't restrict him too much and then ask him to steady up a bit. Hang on tight and hope for the best is the attitude over the first few fences with these horses!

The right equipment

Always remember that safety and prevention of accidents must be your number one priority. A crash hat of the right size, correctly adjusted for cross-country riding, is a common-sense precaution, as is a back protector. Your horse should wear a breastgirth (or breastplate) and surcingle (or overgirth) to hold the saddle in place. Your horse's legs should be protected, especially if he tends to knock himself when jumping.

The one-day event

Arrival So you are fully prepared and arrive at your one-day event. Now you need to have a clear idea of when and how you are going to ride each phase. If you have been given times, well and good. If not, make sure you know the approximate times you are likely to be in action. Find out where everything is so there is no last-minute panic, and give yourself plenty of time to get ready, especially if it is one of your first events. It is surprising how long it can take to get organized, so try to put out all the things you need for each phase when you arrive.

Walking the course Walking the course is the most important part of the competition and you must consider before you start what you are aiming at. Are you using this as a school for yourself or your horse? If it is the first time for one or other of you, are you set on achieving a clear round rather than going by the quicker routes, or are you going out to win and going to take the risks which this bold approach requires?

Having decided what you hope to achieve, it is then very important to assess the course to see that this is going to be sensible and possible. It is no good carefully preparing yourself and your horse only to be confronted with a totally unsuitable course which is too demanding for you or the horse's degree of training, or has badly built fences with false groundlines and near-impossible stridings to negotiate safely. Most courses are, however, beautifully built, and in the right class you should find just enough to make you and your horse think but enjoy yourselves.

Deciding how to cope Decide which are the most tiring parts of the course and where they come. Look at the ground—is it soft and muddy? Galloping too fast in such going is the best way to strain a tendon. Is it very hard and, if it rains, will it become greasy? Do you have to jump from light into dark? Horses' eyes take a bit of time to adjust to this so it is vital to approach these fences steadily to give your horse time to focus. What are the take-offs and landings like and where is the best place to jump? Have you got fences with options?—perhaps an in-and-out or a big bold parallel—which would suit your horse best? A bold jumper could take the spread, but if you are worried by this the slower but smaller route may be more suitable for your green or shorter-striding horse.

Have you got water to jump into? Your training sessions should have ensured that this presents no problem so long as you approach with determination, but not too fast so that the 'drag' of the water on landing makes your horse stumble.

Have you got fences with ditches under? Remember to ride forward over these and never look down. The same applies to drop fences—always pin your sights on a fixed point well ahead of the jump.

Below Tack and equipment.
1 *Dressage saddle*
2 *Numnahs*
3 *Double bridle*
4 *Girths*
5 *General purpose saddle*
6 *Snaffle bridle*
7 *Surcingle*
8 *Irons and leathers*
9 *Rugs and blankets*
10 *Boots*
11 *Weights*
12 *Headcollar and rope*
13 *Lunge cavesson and line*
14 *Side reins*
15 *Grooming kit*
16 *Studs and screw tap*
17 *Spare set of shoes*
18 *Antibiotic powder*
19 *Scissors and basin*
20 *Bandages*
21 *Thermometer*
22 *Insect repellent*
23 *Feed and water*
24 *Haynet*

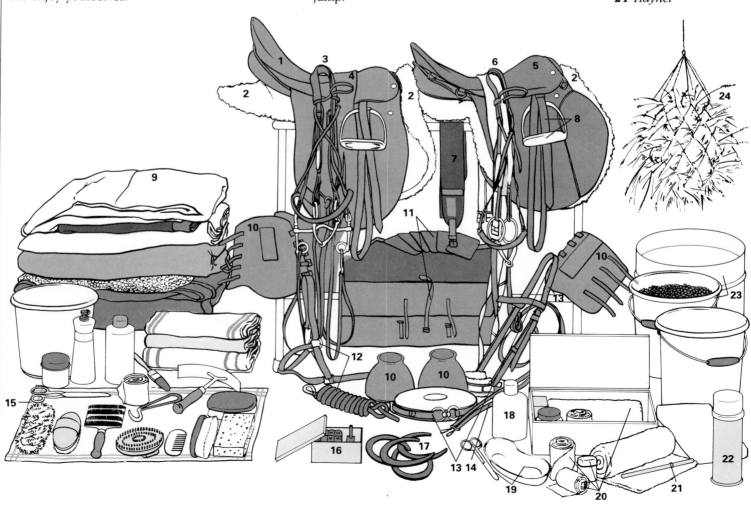

Banks and steps need looking at carefully to ensure there are no roots or rocks in the way. They must be ridden with impulsion if going up, and ridden forward if coming down. You must push your legs forward well as you come down to brace yourself. If you go too slowly the horse will jump down at a very acute angle, making it difficult to stay in the saddle, so make him jump out and forward as much as possible.

Ideally, you should walk the course at least twice if you are new to the sport. The more you compete, the easier it all starts to look.

Show jumping course Your show jumping course is just as important and should also be walked carefully. Give your horse as much room as possible in the arena. As soon as you have jumped one fence, look for the next one. Don't unnerve your horse by swinging round a corner to be confronted by a fence. Don't start before the bell or miss your start and finish markers. Warm your horse up carefully, but don't overface him in the collecting ring. Have someone ready to help you with the practice jump. It is better to jump a few practice fences smaller than ones at home as, with all the previous travelling and the excitements of an event going on, your horse may be a little distracted

and you don't want to undermine his confidence by giving him a fright just before you enter the arena.

Care of horse at the event Look after your horse carefully at the event—keep him warm if it's cold and avoid letting him get chilled after your cross-country. If it is very hot keep him as cool as possible and offer him plenty of short drinks up to two hours before you go cross-country. He should not eat anything for a couple of hours before galloping, but it may be possible to give him a small feed in between your different phases, if there is time. If you are using studs allow plenty of time to put these in—it's a good idea to thoroughly clean out the holes the day before and either re-plug them with oiled cotton wool or put in the very small road studs which can be quickly exchanged for anything else. Do not leave your horse standing for too long in the box if you have very big studs in, as it might be uncomfortable for him.

Have your wash-down bucket, sponge, scraper, water and sweat rug all ready for your return after the cross-country and see him comfortable and settled before you do anything else. Once he has stopped blowing he can have a small drink and then a longer one a bit later, plus hay or a feed. He will be just as thirsty and hungry as you, but don't

Below Clothing and equipment for the event ride.

1 *Dark jacket for show jumping and dressage*
2 *Breeches (two pairs at least)*
3 *Tail coat (advanced)*
4 *Stock or tie and pin*
5 *Shirt and cross-country jumper*
6 *Velvet cap*
7 *Top hat (advanced)*
8 *Skull cap and silk*
9 *Back protector*
10 *Boots*
11 *Spurs*
12 *Towels*
13 *Food and drink*
14 *First aid kit*
15 *Sewing box (for loose buttons etc.)*
16 *Whip (you may also need a schooling whip and lunge whip.)*

(For novice events a tweed jacket is correct for show jumping and dressage.)

give him too big a feed straightaway. A small one followed by a bigger feed later is best.

The horse's legs will need care after an event. If the going has been very hard, something cool on the legs will be good, such as cooling lotion or cold water bandages. If the ground has been soft, something like kaolin will be soothing. Any cuts should be bathed and some wound powder or spray applied. Remember to remove the studs.

Further progress
When you have successfully completed several one-day events you will feel confident enough to progress further. Remember, though, that everything gets a bit more advanced, so, when preparing for bigger events, don't neglect one phase in preference to another. It is always a good idea to aim for a higher standard at home than you are ready for in competition, as horses rarely go as well away from home. It also prevents boredom if you are trying different things and helps to ensure that horse and rider are progressing all the time.

Steeplechase If you decide it would be fun to do a small three-day event, there are some more things to be considered. Riding over steeplechase fences requires a sharp eye and firm grip. Because of the extra speed you must hold onto your horse more over the fences and ride harder into them. Sometimes it might be possible to box to a training stable and get some advice on how to ride forward and hold the horse at speed over steeplechase fences. If not, watch a few steeplechases on the racecourse or on television and study how the jockeys sit down and ride into their fences, but always have a safe, firm contact on the reins and push their legs forward over the fences. You won't want to ride as fast or with nearly such short stirrups, but you can get the idea.

Roads and tracks The endurance phase, roads and tracks, needs little practice, but it is worth measuring out a kilometre somewhere and allowing the horse to trot or canter quietly so that you can complete this in four minutes. This is roughly the speed required on this phase, though you may find this varies depending on the terrain. In the competition there will be kilometre markers. You will need to work out which courses have good ground that you can do in, say, three minutes, and which include a steep hill or bad going which would need about five minutes to complete.

Weights You may have to carry weights in a three-day event, and will in anything over novice standard, so it's a good idea to practise with these if necessary. The minimum weight is 165 lb or 75 kg. If you need a weight cloth it must be correctly fitted; the weight should be evenly distributed on either side, with more in the front, nearer the horse's centre of gravity. As you put the weight cloth on, preferably over a cloth or numnah, make sure that it is well pulled up into the front arch of

the saddle to prevent pressure on the withers. Check this again as you tighten the girths because pressure here could affect the horse's jumping or way of going if it becomes uncomfortable.

Extra work for three-day event If you are training for a three-day event, your horse will need more fast work depending on his type and the state of his wind, and may need a gallop or three-quarter-speed canter once a week, or twice every ten days, in the last two to three weeks before the event. The distance should never exceed more than 1¼ miles (2 km); the first two-thirds should be at a strong but steady canter, pushing on to a fast gallop for the last part before pulling up quietly back to walk. Start by doing less than a mile and gradually ask a little more, but whatever you do don't overdo the galloping. Some horses need hardly any, others need a lot, but try to get your horse to canter in a relaxed way by sitting very quietly and not fighting him, so that he settles quickly and gets into a good flowing rhythm. Listen to his breathing on coming

Above left *You must learn to ride like a jump jockey for the steeplechase phase. Unless you can gallop on at all fences, valuable seconds will be lost. You should walk the course at least twice, once at the time when you will be on the course, to see where the sun will be and if it will affect your approach.*

Below left *Banks and steps are often negotiated at all levels of eventing. Study them carefully and then ride them with impulsion going up and while coming down encourage the horse to jump out and forward.*

back to walk and see how long it takes for him to stop blowing and settle back to normal. The recovery rate should not be more than about ten minutes if he is fit and not being overstretched.

In between continuing the flat work and doing a little jumping the horse should be going out two to three times a week on long hacks to prepare him for the three-day event. He should be trotting and cantering quite some distance in between periods of walking and be out for 1½ to 2½ hours overall, depending on the countryside.

The vet's inspection For the veterinary inspection prepare your horse so that he looks the part; in a bridle and, in most cases, plaited. Find out from other competitors what is normal for the standard of event. Remember to stand in front of your horse whilst the Ground Jury inspect him and then walk out and trot smartly back with the horse's head free on a loose rein. Before running up in front of the jury, make sure the horse is well walked to remove any stiffness from the journey and, especially at the

final inspection, really loosen him up by going for a short hack to get the stiffness out of him and then keep him walking until after the inspection—many horses fail the inspection because they have not been sufficiently loosened up beforehand.

Briefing Ask questions if you are not sure about anything. The briefing is normally followed by inspection of the roads and tracks. Get into the front of a vehicle if you can and make sure you have your map of the course with you so that you can follow where you are going, note down the good and bad patches on the course, and particularly note where any compulsory flags are sited. Check very thoroughly to see how many there are—many first-timers come to grief here by missing one on the way, resulting in elimination. The steeplechase is usually walked on foot but take note of any sharp corners that may be difficult to ride at speed and note where the finish is if the course is not roped— it is easy to go the wrong side if there is nothing keeping you on course.

Above The show jumping course should be walked with as much care as the cross-country and steeplechase courses, and your track planned to give your horse as much room as possible. Be careful not to unnerve your horse, especially if entering an arena with many distracting sights and sounds. As you progress, courses will become bigger and more complicated, with more combination fences.

Dressage For the dressage turn yourselves out immaculately. There may be more spectators and a more tense atmosphere than usual and your horse might need extra work to keep him calm. Afterwards take him for a short gallop to clear his wind—about ¼ mile (400 m) at a steeplechase speed is usually suitable.

Cross-country preparation Prepare everything for the cross-country and thoroughly brief your helper about what he or she is to do. Work out a plan on when to feed your horse, and take out his haynet and water bucket, etc., depending on your given start time. Allow yourself enough time to get tacked-up and weighed-out so that you can arrive at the start about five minutes before your time to go.

Don't forget to start your stopwatch as you set off and check your times, which should be strapped to your arm as a reminder. On the steeplechase, ride steadily into your first fence so that the horse has time to settle into his stride. Come in to the ten-minute halt about ten minutes early to give you more time to refresh your horse with the least amount of fuss. If it is very hot, sponge lots of water onto his throat, chest and head, and up between his hind legs, but be careful not to put cold water over his back while he is still hot as it may cause muscle spasms.

Keep checking your time and if you have taken the saddle off, allow five minutes to get ready and mount. Wake your horse up a bit before going into the start box.

Last day On the last day allow time to unstiffen the horse before the vet's inspection and the show jumping. Really loosen him up without overtiring him and start over some small fences. All horses are bound to be fairly stiff after the cross-country and must be given a chance to get going well before the jumping test.

At the finish look back on how it all went and think where you made mistakes and what to concentrate on next time. There is always room for improvement and it is only by aiming higher that you can eventually get somewhere near the top!

Conclusion

Hopefully, at whatever level of eventing you are aiming, you will get a lot of enjoyment from the sport. Its all-round nature makes it suitable for more people than many other equestrian activities, and whether you do it full or part-time makes no difference to the fun that can be had by participating. Some countries have more events than others and Britain probably offers the best opportunity to train and compete, whilst Australia possibly has the least. Yet both countries have won numerous medals and this only goes to show that determination is the all-important factor. This along with hard work and some luck, should combine to give you years of fun and excitement.

Far left When jumping into water take care not to enter too quickly. Even very shallow water exerts considerable drag on the horse's legs and could cause it to lose its balance. It is wise to have tested the depth of any water when walking the course, to avoid testing the temperature when riding it!

Left When jumping down steps, sit well back and push the horse forward. Do not be panicked by the scientific approach of some riders; stick to basics and keep it simple. Combination fences need steady accurate riding while sloping or wider fences need greater impulsion.

Below Turn yourself out immaculately for your dressage test, at whatever level. Never allow yourself to get worked up, however naughty the horse is. Experiment with your riding-in, perhaps lungeing or just walking about quietly before you start.

6

SHOW JUMPING

What is show jumping? Extensive exposure and increasing popularity of the sport over the past thirty years leaves few lost for an answer. It is, of course, competitive precision riding over a set course of obstacles which, although solid in appearance, are easily knocked down. Each knock or refusal incurs faults which are marked against the competitor. The winner is the rider who can jump the course clear, and then jump at least part of it again, raised and this time against the clock, to be clear and faster than any other rider.

Built to jump

Before training begins it is essential to have the right type of horse for the job. A show jumper needs to be an athlete, much more so than does a racehorse. To do his job, the show jumper has to use slow but powerful movements and he must have basically good conformation. Try not to buy exceptionally big or exceptionally small horses. There have been brilliant horses of unusual sizes, but generally the well-built 15.3hh–16.3hh are best. The actual height in inches or metres is irrelevant, but between these heights the overall size of the horse is ideally suited to the average-size rider. If they eventually upgrade, they will cope when they have to jump the big courses consistently. So many good little horses with tremendous guts and talent sail through Foxhunter or Green hunter classes, but when they have to jump the bigger heights and spreads every day, things start to go wrong. The horse usually starts to stop and becomes useless for competition, having absorbed somebody's time and training for at least twelve months, sometimes much longer.

Equally, the extra large horse is very difficult to ride; all his strides are naturally long and normally a very big horse does not mature physically until he is about eight years old. In the meantime the rider has struggled to hold him together, needing a very strong leg and seat to get real impulsion. Of course,

we have seen Stroller and Dundrum in the 'mighty midget' class, and many of the German horses have been 18hh or more, but these big horses don't last long. They seem to fade out very quickly with leg or back troubles. Rather like very big people, they tend to strain themselves easily. Temperament, too, is tremendously important. A horse and rider must suit each other.

Other abnormalities, such as bad movement, head shaking, big joints, legs or feet turning in or out, excessive fatness or thinness are all to be avoided, within reason. They are potential physical breakdown points as the horse becomes stronger and more advanced in his career, and greater demands are made on his body. For a jumper to have the scope eventually to jump high and wide he must have a good shoulder. Look at your prospective purchase sideways on, and mentally put a saddle on him; if his neck starts at the front edge of the saddle flap he has got no shoulder. You should see almost as much length from the saddle to the beginning of the neck as in the neck itself. A horse should look as three equal parts—front, middle and back. If he looks unbalanced standing still, he probably will be when he is moving.

Early training

A good time to start training a horse for show jumping is at five or six years old, when he has been well broken and perhaps done some hunting or a few shows as a show horse or working hunter. He has seen a bit of life, been in and out of horse boxes, been in the company of other horses, been to different places to work and generally grown up a little.

Beginning at the end and working back, the international Grade 'A' horse will have spent at least three and probably four years reaching that level and, having reached it, should go on in top competitions for anything up to eight or ten years, barring accidents and unsoundness. It is therefore

necessary to find your prospective show jumper and begin training or re-training at about six years old. If the horse has been started correctly, any rider will get the right reaction, but it is up to you to trim up its basic training to suit you personally.

In the first year he will do the basic flat work which is going to be the backbone of everything he does in later life. Call it flat work, ground work gymnastics or dressage, what it develops in the horse is a complete understanding of the rider's commands, and the physical ability to perform them as an athlete, with suppleness, power, balance and cadence. In layman's terms, it puts a gearbox into the horse and finely tunes all the controls to give instant responses.

This initial training must bring body and mind together, beginning with the horse going very freely forwards, more in trot than any other pace. The trot involves enough acceleration to generate plenty of activity in the horse, but not so much as to be difficult to control and, as an even two-time pace with no positive left or right, any change of direction is simplified. In rising trot, of course, the rider must sit on a given, consistent diagonal, usually when the inside hind leg is on the ground, to assist the balance of the horse in turning. If you ride for any distance on roads, or across country, change the diagonal every few hundred yards to prevent the horse and you from becoming 'one-sided'.

Daily schooling
Daily work should consist of an interesting variety of exercises at walk and trot, and later canter, involving circles of about 20 metres (20 yds) to begin with, down to 5 metres (5 yds) as the horse becomes more balanced, changes of cirection and transitions, all ridden with very positive leg aids to build up the understanding of the horse to these aids and the quickness of his mind to put them into action.

From Day One, eight to ten poles can be put on the ground at 4-foot (120 cm) or 10-foot (300 cm) intervals and the horse ridden over them, still at the trot. This begins to teach him to be careful. No horse likes to tread on a pole; it is awkward and uncomfortable and so is much easier to make the effort to step over them. The line of poles also accustoms the horse to seeing the long line which will, in time, prepare him for combinations. In the line with poles at 10 feet (300 cm), every second pole can be raised to about 2 feet (60 cm) and then the operation can be ridden out of at a canter, or sometimes cantered all the way. If the horse rushes, come back to trot and circle away after two poles. This brings in an even stride and small jumps, building up a 'flow' pattern; it also lowers the horse's head, and rounds his back as the continuity of the line keeps him looking down and forwards.

When schooling a horse, don't follow exactly the same routine every day. As you begin to jump small fences, the object all the time is to perfect the horse's technique; every jump should see maximum effect for minimum effort. Small solid fences, such as logs and rails, or sleepers, or anything of that nature, make excellent schooling fences. The horse respects them and learns to be careful—self-preservation is pretty high in most horses' minds. If you are lucky and live near to land where you can jump hedges and ditches this makes good variety for both you and your horse. An introduction to every possible kind of low-level jump is invaluable, so that jumping black fences, ditches, ups-and-downs, streams and fords all becomes part of the daily routine. Then, when the horse meets a water jump or a bank for the first time he is not shattered by the experience. These days more and more 'Derby'-type courses, with semi-'natural' and permanent fences, are springing up and a horse needs educating over similar fences before tackling them in a ring.

Overleaf *The show jumper's early education should prepare him for the more permanent, fixed-type of obstacle, which he will meet in his career.*

Below *A horse and rider tackle a double combination fence of an upright followed by a parallel and depict the correct ways and some faults of show jumping.*

1 *The rider keeps a good seat as the horse lands, to ride it forward to the parallel.*

2 *The strain of landing is taken by the fore-foot and tendons and ligaments below the knee.*

3 *The rider prepares his horse for take-off, maintaining an even contact with its mouth.*

How many times a week should you school a horse? Think how you would feel if you were in training; maybe you should train every day for a while, settling into a pattern which produces an active, attentive horse. Intersperse serious work days with some other more relaxing work off the premises.

Riders make problems

In general at this stage of training I find the most common problem is bad mouths on the horses, mostly created by riders with too strong hands in relation to their seat and legs, and a poor way of going on the flat between fences. These riders tend to drive rather than ride their horses. All courses are built around a standard measurement representing the average correct horse's non-jumping strides. Therefore, it is most sensible to achieve, as far as you can, a way of covering the ground that fits the standard pattern. Distances between fences will then be normal and easy for you and your horse. The cure for overstrong hands? Easy—take them away! Pulling horses are made by pulling riders, with very few exceptions. Riding with both reins in one hand is a help to lighten hands, and riding without stirrups strengthens your seat, corrects your balance and improves your technique and feel. Ride circles at slow paces in good balance and on a very light contact and retrain yourself to this system. The horse will be much more willing and more happy to go sensibly.

If possible, get someone responsible and knowledgeable to lunge your horse with you on him, in walk, trot and canter, so that you can ride without reins and feel how the horse responds to your body weight and leg pressures. You can also jump in this way, as it allows the horse complete freedom of head and neck and he can jump in the way he prefers.

When jumping, it is the rider's job to set up the correct pace and direction and create enough impulsion for the horse to jump whatever is asked of him. It is the horse's job to take over just in front of the fence and jump it. The rider must take great pains to ensure that his weight is never hindering the horse over a jump; too far forwards before and during take-off buries the horse, sends him onto his forehand and will inevitably cause many faults and probably stops. Too far back hinders the hindquarters and forces the horse to lift his head and hollow his back to counter balance; this will make landing difficult and very heavy and, in extreme cases, jar the horse's back, again causing him enough discomfort to make him want to stop and learn to refuse. Similarly, any shift of weight to either side may hinder the horse's balance.

At the moment of jumping, the rider simply folds at the waist and goes with the horse's movement as naturally as possible. Beginners should be taught to find their own balance by holding the mane or a short neck strap as they jump small fences on a loose rein.

The first show

Having completed a few months of basic training you are ready for, and need, some shows to see how you are progressing. No amount of work at home can bring the horse or rider beyond a certain stage. Both need the exposure to the show and its atmosphere to seek out the chinks in the armour.

Travelling costs money and taxes the horse even more than the jumping or work side of his life. Consider your own energies too; if you're riding and driving, work out your day so that you are not going to be totally exhausted. Give yourself at least one hour before your event to allow time to unload and prepare the horse, bearing in mind that after the effort and expense involved in going to a show, it is wise to be a little early rather than late.

If the horse is very new to shows, get him out of the box straightaway, tack him up and ride him in among the other horses; let him have a good look

4 As the horse takes off, the rider has shifted too much weight to one side.

5 The rider's weight is still unbalanced. He should look straight between the horse's ears.

6 The rider allows with his hands so that the horse can stretch to make the parallel.

around and realize that he is in a new place. Don't expect him to give his full attention to you until he has had a chance to see all the new sights. This is the time when a bit of nervous tension can be quietly released without making either party upset in any way. When the horse is settled, return to your box and finish dressing yourself, ready for the competition. Put in studs, if necessary, and boots, over-reach boots or any other tack on the horse. Always carry spares of everything in the box, including extra rugs.

Walking the course If possible, have someone travelling with you to help you, and ask him or her to hold your horse, preferably walking quietly round, while you walk the course. This is most important as it is your one and only chance to see the course at close quarters and weigh up the fences one by one. Note any fences going directly away from the collecting ring as the inexperienced horse in particular will always hang towards the

way out. Doubles and combinations always cause the novice horse to have an extra look and you may need to attack and ride a little more strongly towards these—but take care not to over-ride and destroy the rhythm. Take particular note of any slope in the ground; the horse will always run with the ground slightly, to either side, and obviously uphill requires a bit more push, downhill a bit less, but impulsion must be maintained whatever.

Remember it is the rider's job to work out all the parts of the course on the flat, and to present the horse at the fences in the easiest possible manner. As courses get bigger in the higher grades this becomes very critical, and must be calculated to within inches to arrive correctly at each fence. Whatever happens in the ring, try to ride as normally as possible. Horses are quick to cotton-on if they get away with things in the ring.

General hints If you have a novice horse, try to make shows a part of his routine which is neither a

Above left *Double bridles and draw reins are useful aids, but only in the experienced hands of riders like Britain's David Broome.*

Top *The rider is hampered by using too long a stirrup; the leg is too far back and the body has tilted forwards.*

Centre *Because the rider's seat is too far back, she is unable to ride the horse forward to the next fence.*

Bottom *Left in the 'back seat', the rider is likely to give the horse a nasty jolt in the mouth.*

wildly exciting nor shocking experience, but one that he can look forward to and enjoy as a change from the home routine. Teach him his manners, in just the same way you do at home. Don't let him run your life by being a maniac or a pest; a disciplined horse is a happy horse! Above all, give him time on the flat to warm up properly. Don't over-jump at the practice fence, either in height or in the number of jumps you ask him to make. Six to ten pops is plenty if all is going well.

Enter the arena boldly, and have a good canter round before starting the course. If you are ready at the collecting ring to go in quickly, you give yourself that few extra moments before starting your round, and you won't harass the poor steward, who is probably a volunteer for the job and will stand for many hours whatever the weather calling horses into the ring. Shows are run for competitors, and competitors are essential to shows, so helpfulness and co-operation is the name of the game.

Further training

Reflections on the way home? Perhaps the horse was too fresh; maybe the presence of all the other horses upset him—there may be all kinds of things which are only going to be sorted out by having a few more outings. Often the filling of the show fences is more than the horses are used to and this makes them jump rather stickily. Fill in fences at home as much as possible while the horse is a novice and after a half-dozen shows he will become quite used to this.

In between these early shows keep up the gymnastic work in just the same way as before. If it becomes clear that the horse needs more work on a certain type of fence then construct a small edition of that fence and practise him over it, but all at slow paces, allowing the horse to think for himself. Use placing poles, about 5–7 feet (150–215 cm) from the base of the fence, to help the horse and rider 'arrive' consistently correctly, and

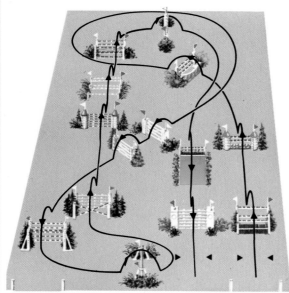

Above *The layout of a typical show jumping course indicating a route for a round. The smallest course should have space for at least 10 jumps placed at least 60 ft (18 m) apart. Each is numbered and may carry white and red flags for left* *and right respectively. The basic jumps are uprights, such as parallel poles and the wall, which can exceed 6 ft (1.8 m) high and spreads of poles, which can often be as wide as the uprights are high. Double or treble combinations, are one or two paces apart.*

Above *The experienced rider, while allowing with the hands and moving the upper body forward with the horse, keeps her seat firm. The rider's aim always should be to maintain a position to ride the horse forward with impulsion.*

Above right *At a competition a horse must get used to the practice area, where both horse and rider will have to concentrate on the job in hand, and be aware of, but not distracted by, other competitors.*

then take them out. You will probably find that you have not nearly as much control as you thought you had, so try different bits at home to find out what the horse readily accepts. One of the most handy snaffle bits for jumping is the French bridoon. This bit is light in weight, considerably stronger than any plain snaffle, but not severe. Most horses accept it, and it has the effect of keeping the poll flexed and the horse balanced and supple if he tends to be a little over-strong. Not many horses go well in pelhams for jumping, as the action of these bits is rather blunt and usually too strong, with a loss of 'give' from the horse. Consequently he may fight the bit and not go forwards at all, or go in a jerky, stop-start fashion. Corners tend to get 'lumpy' for the same reason and you generally lose the flow and rhythm of the horse's stride through the course. Bridling a horse both to the job in hand and to the rider is a long and complex subject, but basically horses are overbridled, rather than under. If you have a real problem, don't be afraid to get professional help. For starters, you can take everything in the way of martingales and bits off the horse, and begin again with a simple snaffle, trying to analyse the cause of the problem. Does the horse have a 'mouth'? Has he wolf teeth? Is he resenting the bit, frightened, or just not responding? Has he had a running martingale set too short, pulling down into his mouth and bruising it? Does the

rider use enough leg and does the horse respond to the leg? Is the horse schooled enough? Look for all the simple, logical things which will add up to make a picture of the horse's background and help you to work out what must happen next to correct the problem.

Managing the show jumper

The jumping horse's performance will be greatly affected by his general well-being. The first thing one is always taught is 'no foot, no horse'. Take particular care to have the feet regularly shod by a good farrier who understands jumping.

Take care to fill the stud holes with oiled cotton wool whenever the studs are not in. Small stones can work their way into the holes and right into the foot, setting up an abscess—not to mention the job you will have cleaning the holes out everytime!

Hard ground is a greater enemy to the jumper than soft going. Hard going creates concussion of the joints and bones of the leg, especially the forelegs. Light, firm, even bandaging over gamgee or other wadding helps for work and, at night, a bandage on top of a pad well soaked with a solution of vinegar and water, or Radiol and water, loosely wrapped will make a comfortable 'bedsock'.

Feeding and balancing the food to the work is something you find out about as you go along, but basically you should feed a show jumper like a

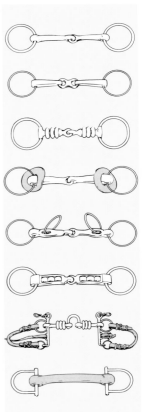

prize fighter. Each horse will be slightly different. Remember that the jumpers usually have a long day at a show, so work out feed times at home to fit in, so that on show days regular times are not too disrupted. Obviously horses don't want a big feed or drink just before competing, but equally they don't want to be starved for hours.

Horses need the revitalisation of a holiday at grass so try to give two months a year, one in spring and another in the autumn.

Preparing for the top

You may ask yourself if, after all this training and painstaking education, there is any chance of reaching top-level competition. The answer is, quite simply, yes. But not by chance, or even by winning a few classes. The road to the top in show jumping gets narrower towards the end and only a few go on to the international level. The main reasons are that not only will you have to devote almost all your time to the job, but your costs will escalate as you start to travel further afield, and you will need a great deal of natural talent as well as dedication. Of course, one really good horse that wins a fair bit and thereby pays his way would be marvellous, but most people who are really into show jumping have more than one horse.

In fact, work-wise, horses can be run in units of four. One person can look after, feed, groom and travel four horses. One rider can work and compete with four horses, normally in different classes, and sometimes one or two horses have to be sold to bring on new youngsters. This way it is possible, with careful selection when buying, to pay for your hobby. You are unlikely to make a profit, unless you happen on a really brilliant horse and take a big offer for him. And you may lose, for horses are increasingly expensive.

Equipment is expensive initially, but if it is looked after carefully and kept in a dry room it should last a very long time. Transport is probably the greatest expense besides the horse himself, and of course vehicles do wear out and lose their value, but again careful handling, with good maintenance of paintwork and inside woodwork, will greatly extend the life of any box or trailer.

Always a challenge

Whether you ride purely for pleasure or aspire to greater things, such as going abroad with your country's colours sewn on your jacket, horses in any form are a tremendous challenge, great fun, full of ups and downs, and can provide an endless form of healthy outdoor activity. As people are by nature inventive and competitive, so long as there are horses and riders on this earth, someone will keep having new ideas and adding to our range of competitions.

Above Types of bits described from top to bottom. The vulcanised jointed snaffle suits a young horse as rubber creates more sensitivity than metal. The French Bridoon is a soft, lightweight bit for a good mouth. The roller snaffle is fairly heavy, and suits a bad, hard mouth. The German snaffle is a hollow, light-weight bit that only works on a good mouth. The Cornish, or Scorrier, is very tough, but not severe, and is useful for strong horses. The Magennis works on the cheeks and lips to create more feeling in an insensitive mouth. The Hannoverian Pelham, which can be fixed or jointed, prevents the horse holding the bit and the tongue port helps stop the tongue going over it. The rubber snaffle is excellent for sensitive but strong mouths.

7

DRESSAGE

The ranks of dressage riders are growing rapidly. More and more people are awakening to the challenge of establishing that partnership with their horses whereby they can teach them to use their bodies with increasing grace, power and suppleness—to become equine gymnasts.

The dressage horse
When choosing a horse to train for dressage the first and most vital consideration is character. The training of your horse is a task of years and it is better to hold an affection and respect for him, rather than just put up with him as the basic tool of your sport. A harmonic relationship is the basis of successful dressage and it is important to like and understand one's partner's mind.

The characteristics to look for are boldness, energy, interest in work and a little bit of devil; an impishness which makes the horse want to show off. Those to avoid are stupidity, meanness, laziness, and a horse which panics. One can get a pretty good impression by looking at the head and, in particular, the shape and the way the eyes and ears move; be wary if these are small and move nervously and quickly. An even better impression can be built up by judging the horse's reactions to situations: how he behaves when faced with frightening objects, a strange obstacle to jump or a strong demand from the rider/lunger.

The most important practical characteristic for really good dressage horses is athletic ability. They must have the power in the hindquarters to enable them to spring, they must have freedom of the shoulders which enables them to take long strides and their bodies must have that suppleness, looseness and co-ordination which will enable them to work like gymnasts.

Good natural paces are a tremendous asset and probably vital for a really top horse, although my own Grand Prix horse lacks them: turned loose, and since a youngster, he has walked with strides that miss overtracking by 6 inches (15 cm) or more, and he spends much of his time when loose in a disunited or four-time canter—but he has got

character and that vital athletic ability with natural power.

Another vital practical asset is that the horse has the make and shape which will help rather than hinder the learning of dressage and make it more likely that he will stay sound. It is rather difficult to define this, as good dressage horses do come in all shapes; plenty would not win in the show-ring, and I think that character and athletic ability are more important than perfect conformation. However, conformation should be robust and any defect must be carefully examined, with care taken that the usual bad results of such a shape are not apparent.

The importance of dressage in Europe has led to the specific breeding of suitable horses. Thus very large numbers of Hannoverians, Trakehners, Holsteins, Swedish Warm-bloods, Danish Warm-bloods, Dutch Warm-bloods, Selle Français, etc., have satisfactory talent for dressage. The Thoroughbred is a much trickier proposition. Those with great athletic ability are sold for fortunes as potential Derby or Gold Cup winners. They are bred specifically for the race-track not the dressage arena, thus they gallop well but usually find it difficult to compress their bodies and spirits into collection; most want to beat other horses rather than accept the discipline of becoming equine gymnasts. Nearly every rider, however, if asked to select his favourite breed/type of horse—character, athletic ability and conformation being equal—would choose a Thoroughbred above all others.

The basic material needed for dressage is thus found in larger numbers amongst the breeds of Europe and, although the English-speaking nations rarely breed horses specifically for dressage, Thoroughbreds and cross-breds can be found with the natural ability to be trained for this sport.

Early training
The training of a dressage horse starts on the same lines as for any other competition horse and it is only after about two years that it becomes more specialized.

The first training stage is similar, but, as one of

156

the aims of dressage is maintenance and development of natural paces, I usually spend at least six weeks lungeing so that the horse has the strength, balance and mental stability to take the rider's weight without spoiling his paces. I am a tremendous believer in lungeing (when well done) and my young horses have two or three sessions a week; even my advanced horse still has one lunge session a week.

After the young horse has accepted a rider upon his back and is reasonably under control in an enclosed space, the main aims of the next stage of training are to develop his strength, to encourage his desire to go forward and to win over his mind so that he obeys the aids and remains calm. Lungeing sessions, hacking out (if possible in the company of another horse) and very occasionally short schooling sessions help to achieve these objectives.

A vital consideration in dressage is that the work should be fun for the horse. This means that the horse must be strong enough to do the work—physical pain or tiredness will soon make him fed up—hence the time spent hacking out, going up and down hills to make him fit.

It is also no fun for the horse if he has not developed the desire to go forward, for then he will be subjected to constant thumps from the rider's legs and frequent use of the stick. The development of this desire to go forward starts from when he is first led. I always ask my horses to walk actively beside me (achieved by use of a long stick in my left hand to use on the hindquarters, or by an assistant behind) and never pull them along. It continues on the lunge when the horse must learn to walk and trot with long strides. I like my horses to go forward to the extent that if they went any more they would start to run and take hurried, short strides. To achieve this the lunge circle must be large, ideally about 20 metres (20 yds) in diameter, and the horse must realize that you really mean it when you ask him to go forward. With Thoroughbreds this is easy and usually it is more a matter of calming them down, but with the halfbreds it means being very firm, with perhaps the odd disciplinary session, so that you do not have to nag them continuously.

When riding, try to take little or no contact with the reins so that there is no hindrance to forward momentum; when out hacking, another horse in

Above Skilful lungeing (1) helps a young horse to develop its balance and paces without a rider-burden. It should be encouraged to work long and low (2), stretching its head and neck down, taking a contact with the rider's hands and using its back and quarters. The walk pirouette (3), using only 90° turns for a young horse increases suppleness and encourages lightness in the forehand. The rider can sit to canter (4) as the young horse develops, but should adopt a forward seat if resistance is felt.

Overleaf Even at Grand Prix level, the dressage rider must continue to practise and learn.

Above *The mark of a potential dressage horse is one with a bold, energetic character and with above average athletic ability. It should have powerful hindquarters, freedom of the shoulders and a body that is supple, loose and co-ordinated.*

Overleaf *A full-size dressage arena, laid out for a Grand Prix competition.*

rein contact while allowing him to stretch and encourage the activity of the hindquarters so there is a feeling that his withers and back rise upwards—he is bent like a bow. This has the wonderful effect of stretching those vital top-line muscles, developing their strength and the spring to the trot and canter. I work for this after the horse has learned to accept the bit through use of unilateral half-halts first to the left then to the right and by use of serpentines and figures of eight.

As the horse becomes more supple, strong and balanced, begin a little sitting trot, and sitting in the saddle at the canter. The moment you feel any stiffening or hollowing, return to the rising position. It is vital that the horse's back remains rounded and supple, and impatient and/or stiff, poor riders soon destroy these vital ingredients.

The next stage is to start a little lateral work (moving sideways as well as forwards). I use leg yielding a good deal (for definitions of movements, see national or FEI—Fédération Equestre Internationale—rule books), but others start with a turn on the forehand. As soon as the horse can remain on the bit in all his work, can complete smooth upward and downward transitions, is pretty straight at the canter, can lengthen his stride slightly at the trot and maintain the correct bend on curved lines in his work, you can consider the possibility of competing in the easiest tests. Ideally, aim to be working at home at a stage higher than that at which you want to compete. Note that to be 'on the bit' the horse must not only accept the bit but maintain a steady, correct head position (that is, face just in front of the vertical) with the hindlegs active and well engaged.

front helps this aim, as do plenty of canters and strong trots on good going.

Seven objectives

When the horse is beginning to muscle up, goes forward happily and is calm (usually after about three months) you could start the odd schooling session.

There are seven main objectives to these early schooling sessions, usually covered in about nine months, before the horse is ready to compete. They are impulsion, rhythm (and balance), suppleness, straightness, acceptance of the bit, submission, and development of the paces.

Developing the horse

As the horse starts to accept the bit, use more circles, usually not less than 20 metres (20 yds) in diameter, serpentines and transitions, all the time aiming to develop suppleness and straightness in particular.

I am a tremendous believer in asking a horse to work long and low—that is, to stretch his head and neck down towards the ground; but when doing so he must not fall onto his forehand. Maintain the

The first competition

The first step towards competing is to choose an event where there are good riding-in facilities, a relatively peaceful atmosphere and the going is good. Good going is vital—a slippery surface or, worse, a very hard one, have detrimental effects on the paces. The former makes it difficult to maintain rhythm, the latter jars the horse, stiffens his back and, over a prolonged period, leads to a loss of spring and power to the trot and canter.

The second step is to make efficient preparations. The test must be learned, even if it is to be commanded on the day, and it is not only a matter of mastering the order of the movements (best done by trotting around on your own feet in a miniature arena!) but also thinking carefully how to ride each movement and when to start preparing for it.

We must learn to ride a positive test at home—first thinking it through in the mind and then getting on the horse and practising it in an arena. Do not run through the whole test too often, otherwise the horse will start to anticipate, but sections can be practised out of order.

The other part of the preparations is to collect together all the gear for the horse, making sure you know what tack is permissible in the rules: saddle, snaffle bridle, brushing boots, studs if the arena is on grass, travelling equipment, lungeing tack if appropriate; and for the rider, hunt cap or bowler, hacking jacket, breeches, tie or hunting tie (stock), boots, gloves, dressage whip and spurs if necessary.

On the day of the show it is vital to be organized well enough to run to time. I like to arrive at the event half an hour or forty-five minutes before I want to start riding my horse. With a young horse I would allow about an hour before the test, lungeing him for some time, followed by walking him around all the sights and probably only starting serious riding-in fifteen or twenty minutes before the test. It is a good idea to establish from the ring steward whether or not the class is running to time.

Just before going into the arena, brushing boots have to be removed and it is time to take a couple of deep breaths to help rid muscles of any competitive tensions. Then it is into the arena, where the aims are to concentrate on retaining a great communication with the horse, to ride positively, *but to enjoy it*, because things will go much better than if you are cursing yourself or your horse for making a mistake or trying to get the test over and done with as quickly as possible.

A test for novices

Tests vary from country to country, but Britain's Novice 12 is pretty typical of the type of test which one would be happy to use as an introduction to a horse's competitive life, although simpler tests can

An exemplary position: the rider sits with a straight line through his ears, shoulders, hips and heels, eyes forward and hands level.

At the halt the horse should stand immobile and attentive and, unlike this one, square in front.

In medium walk the horse should move forward energetically with even and determined steps.

At free walk the horse is encouraged to stretch forward and down, taking as long strides as possible.

		THE BRITISH HORSE SOCIETY'S DRESSAGE TEST NUMBER 12	Max Marks
1	A	Enter at working trot (sitting)	
	X	Halt. Salute.	
		Proceed at working trot (rising)	10
2	C	Track right	10
3	A	Working trot (sitting) serpentine 3 loops, each loop to go to the side of the arena finishing at C	10
4	M	Working canter right	
	B	Circle right 20m diameter	
	BAE	Working canter	10
5	Between E & H	Half circle right 15m diameter returning to the track between E & K	
	K	Working trot (sitting)	10
6	A	Working trot (rising) serpentine 3 loops, each loop to go to the side of the arena finishing at C	
	C	Working trot (sitting)	10
7	H	Working canter left	
	E	Circle left 20m diameter	
	EAB	Working canter	10
8	Between B & M	Half circle left 15m diameter returning to the track between B & F	
	F	Working trot (sitting)	10
9	A	Working trot (rising)	
	KXM	Change rein and show a few lengthened strides	10
10	C	Medium walk	
	HXF	Change rein at a free walk on a long rein	10
11	F	Working trot (sitting)	
	A	Down centre line	
	G	Halt. Salute.	
		Leave arena at walk on a long rein at A	10
12		General impression, obedience and calmness	10
13		Paces and impulsion	10
14		Position and seat of the rider and correct application of the aids	10
		TOTAL	140

of course be used to help accustom the horse to a competitive atmosphere and the arena itself. Each movement of Novice 12 is discussed below.

Movement 1 Try not to enter the arena until your horse is settled and working with rhythm. This might entail doing a few small circles close to the arena entrance, after the judge has sounded his bell or horn to signal your start. For a first test, aim for a pretty steady working trot with, above all, rhythm and balance. As the horse becomes more trained and experienced, you can ask for a more forward-going trot with more impulsion, but with young horses remember that you should create only as much impulsion as you can control in that small arena. Apply the aids for the halt a long way before X so that it can be achieved smoothly with the horse remaining on the bit and not swinging his hind-quarters. For a high mark the halt should be square—that is, with the pair of forelegs and the pair of hindlegs parallel, but with a young horse which is not yet well balanced, the forelegs together with one hind leg left slightly behind would be acceptable. The horse should remain absolutely still, attentive and on the bit while you salute the judges by collecting all the reins and whip into one hand and lowering the other to the horse's side (or removing your hat if you are a man) to make a quiet bow. After gathering the reins, give the aids to move off, taking care again that the move is smooth and straight. To achieve this it is best to keep your head high, look well ahead and keep the legs firmly around the horse.

Movement 2 As you approach C apply a half-halt to help the horse balance himself for the fairly tight turn. Try to help him keep his rhythm and maintain a slight bend to the inside as you turn. The important aspect of this movement is to show off a good trot, so you need to have impulsion, a good outline and pronounced rhythm.

Movement 3 For good marks in a serpentine it is important to maintain a good working trot, hopefully established in the previous movement. Rhythm is again all-important, as is sufficient impulsion. However, it is important to remember that this impulsion is contained energy rather than speed. It is all too easy to keep applying the driving aids with the result that the horse simply takes quicker strides and becomes more tense and stiff. Those driving aids should result in him taking longer strides, not quicker ones, with greater engagement of the hindquarters. The next important aspect of the serpentine is correct bend—that is, as the horse changes direction over the centre line he changes the slight bend throughout his body from one side to the new inside. Finally, it is important that the figure is ridden fluently and accurately with each of the loops being of equal size.

Movement 4 It is important to make good preparations for the transition into the canter. This

In the collected trot the horse's quarters maintain an energetic impulsion, which makes it more mobile and lighter in the forehand.

In working trot the horse should show itself to be balanced and moving forward with pronounced rhythm in even, elastic steps.

In medium trot the horse should be encouraged to move forward with free and moderately extended steps – not quicker and more tense strides.

In extended trot the horse uses its quarters to generate greater impulsion to lengthen its steps to the utmost.

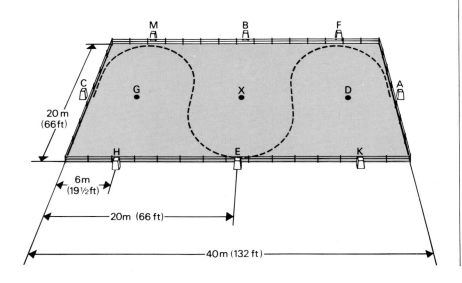

means building up the impulsion with half-halts, establishing the correct bend to the inside to help the horse strike off on the correct leg, and then to apply the aids with tact so that he is not caught unawares by a great kick. The transition must be straight, smooth and clear. The circle must be accurately ridden, the bend must be slightly to the inside and, with the young horse, it will take good use of the aids to keep him balanced, in a good rhythm and with sufficient impulsion. The judge will be watching whether the horse is straight along the long side. It is all too easy for the horse's hindquarters to be further in than his shoulders, so think of the aids for shoulder-in in order to help keep the hindquarters on the track.

Movement 5 To keep the horse balanced in a 15-metre (15 yd) half-circle and to return to the track straight, without the hindquarters falling in, takes positive tactful riding. A few half-halts are needed before turning, just before M, into the half-circle to ensure that the horse is not on his forehand and that his hindquarters are well engaged for the turn.

Do not turn too soon, or you will make the figure too difficult. Again, the bend and rhythm are important, although the horse should be straight from the end of the half-circle to F. With a young horse avoid any counter-canter and return to the track at F. It is very easy to fall into the trot, going heavily onto the forehand in this transition. It takes a good deal of work to keep the horse's hindlegs engaged enough for a clear transition, so that the rhythm of the trot can be established immediately.

Movements 6–8 As above.

Movement 9 At A try to balance the trot as much as possible and start to build up as much impulsion as you can handle safely in order to give the horse the power to lengthen his strides. Never ask for this lengthening, however, until he has taken a few strides on the diagonal, is balanced, and straight. Take care that your driving aids (mainly the legs with a young horse) ease him out into longer strides which are in the same tempo (speed of the rhythm) as the working trot. It is no good just increasing the speed because then the strides

Below A good impression can be formed by looking at a horse's head; the shape and size of his eyes and ears and the way they move.

become quicker and quite often shorter as well. Re-establish the working trot before M, so that the horse is balanced as he returns to the outer track.

Movement 10 Start working for a clear, smooth transition to the walk as soon as you turn into the short side. Once in the walk, keep a very light contact with the reins and try to sit quietly but staying supple so that the rhythm is not upset; but if necessary giving enough encouragement to march forward positively. At H open your fingers on the reins and allow the horse to stretch his head forward and down, and to take as long strides as possible. Do not let him snatch the reins out of your hands or get excited and start to trot. Shortly after X, start to collect the reins gradually so that you are ready to ask for the working trot at F.

Movement 11 Aim for a very calm trot so that you can make a balanced turn, keep straight down the centre line and enable the horse to make a smooth transition into the halt. Allow your horse to establish the halt so that it is as square as possible, but don't interfere once he has settled. Then it is time to salute and to give the horse a pat (which hopefully he will deserve) before leaving the arena on a long rein.

Reward your horse again afterwards—it is rare to be satisfied with a test but it is futile to punish the horse for any faults you both may have made; he will not understand such delayed action.

New movements After the first show there will be much corrective training to be done. Deficiencies in the horse and of the training are likely to have been made obvious. It is very unlikely that the training objectives will have been maintained in the arena; so some hard work will have to go into raising one or more of them to a higher standard. Then new movements will have to be introduced gradually, such as counter-canter, change of leg at the canter, first through trot and then through walk (simple change), and 15-metre (15 yd) circle at the trot and canter. I would usually introduce these movements during the second year of training. At this time, too, I would be beginning to think more about collection, teaching the horse to take shorter, higher strides by greater engagement of the hind-quarters.

Most of the more advanced movements are done in the collected paces, for it is in these that the horse has most power and manoeuvrability. Collection can only be developed gradually, however, as it requires a balanced, supple, submissive, straight horse with great impulsion. It is only achieved by asking for greater engagement of the hindquarters, not by slowing the horse up by pulling on the reins. It is worked for in training through transitions within a pace (working to medium, etc.) and between paces, using small circles, serpentines and shoulder-in.

In the third or fourth year it is time to think of sequence changes (changes every fourth, third and

The canter pirouette requires great collection to keep a three-time rhythm while moving on its quarters.

Changing leg every stride the horse appears to skip, remaining light, calm and straight.

In half-pass the horse moves sideways but always forward and parallel to the arena edges, its outside legs passing in front of the inside legs.

For piaffe the horse must trot on the spot. Each diagonal is raised alternatively with even rhythm and slightly prolonged suspension.

second stride), canter pirouettes—starting with halves and progressing to wholes—zigzags in half-pass (for example, half-pass to the right followed immediately by one to the left and then back to the right). Most horses with adequate talent, good training and health can manage this standard of work. The crunch comes at the final stage. For horses to master the three most advanced movements—flying changes every stride, passage and piaffe—takes talent, a good temperament and skilled training. One-time changes demand fast reactions and good timing from the rider, and great suppleness by the horse to spring from one set of leads to the other. To trot with the elevation of passage takes great power on the part of the horse as well as a special understanding between horse and rider. For the horse to defy instinct and put all his power into trotting on the spot rather than going forward are the extraordinary demands of piaffe.

For Grand Prix work, not only are the movements difficult, each demanding different skills, but it is also necessary to master all of them. For this reason we do not see many Grand Prix horses, and not all those are good ones.

8

HUNTING

Foxhunting as we now know it was born in Britain, but in the Middle Ages, horsemen of Eurasia would ride out with hounds, or trained hawks, in pursuit of duck, birds, hares or deer. The boldest rode after big game: wild bear, lion and leopard, which they stalked and chased over rough hilly country, or through forests, spearing or shooting their quarry at close range. Through hunting, unknown parts of countries and undiscovered routes became known and the world widened.

But the early huntsmen expected their horses to be no more than functional—even expendable in some cases. Through the ages, paintings have shown horrific falls over 'unjumpable' obstacles, demonstrating that there have always been sportsmen with more courage than skill. Doubtless they enjoyed themselves, but what about their horses?

That training is practical, even essential, for *all* horses and riders has always been true. Unfortunately, the hunting man has been known to neglect his own and his horse's education, often regarding the challenging thrills of the unknown, when following the devious and unpredictable route of a fox, as the ultimate in fun which too much 'science' could only spoil.

The rider's responsibility
The rider must take sole responsibility for his mount in the hunting field. Although there are still some people who ride simply in order to hunt, nowadays most people who hunt are there to enjoy the ride. However, it is best to strike a balance between the two, understanding both your horse and the sport. An ignorant or learner-rider can happily survive a day's hunting, relying mainly on courage, grip and luck. He is probably unaware of his horse's discomfort and annoyed if his mount refuses to jump. How much better and more enjoyable it is to ride a trained horse with skill, so that you can cross the country 'as one', overcoming every hazard or obstacle as you meet it. To many a true horseman, there is no greater pleasure than the challenge and excitement of a good day's hunting. There are no records of his performance,

no prizes; but there can be immense enjoyment and satisfaction, as witnessed by the many competition riders, including jockeys, who enjoy a day with hounds.

Some people have a natural tendency to ride competitively in the hunting field, but you soon learn that your *real* opponent is the unknown, and that you must ride wisely to conserve energy. You do not know what lies ahead—a long or a short hunt, over hilly, arduous country or fast, flat stretches. You must also ride according to your horse. If you let excitement transport you, your horse may not be sound next time you want to ride. Most horses, and some riders, are inspired to perform greater feats of courage when hunting, and jump obstacles which they would never dare 'in cold blood'. Indeed, many hunters refuse to leave the ground from the end of one hunting season to the beginning of the next, yet will happily sail over gates and mammoth fences when in exciting company following hounds. With common sense, any reasonably good rider with some courage and a fit horse can complete a hunt.

The hunter
Good hunters come in all shapes and sizes. Age and conformation matter little, but a hunter should have a sensible temperament, be sure-footed and bold. Stamina can be increased with systematic fitness training, whilst speed and extra-talented jumping ability are great assets but not essential. Basically, to be a well-balanced and comfortable ride, the horse must have been well schooled.

Most horses love hunting; the sights and sounds, and the excitement of galloping and jumping with other horses are preferable by far to being confined to a small stable day after day, with little to interest or excite them. A well-schooled and carefully ridden horse will enjoy himself and return home in good, sound condition, while a very placid or jaded horse is often revitalized by hunting, becoming much happier and even displaying talents his rider never knew he possessed. Any horse which is fit and has been well trained to

produce his maximum performance with the minimum effort will complete a day's hunting better than a high-class Thoroughbred type which is unschooled and less fit.

A horse which pulls hard all day, or is unruly, will tire you both and could be a nuisance to others. His good manners are important. He should be prepared to jump whatever you ask (within reason, of course) and you must therefore realize his limitations of ability and endurance. When hunting in open or mainly grass country, like the Midlands of England, where galloping and jumping are the main features, a Thoroughbred-type horse will be more suitable than the heavier cob type, which is better equipped for difficult or hilly terrain.

Not *all* horses are suitable for hunting. Too often a horse which is too young and physically immature is ruined by a hard day's hunting: becoming unsound or at best reluctant. A highly strung animal may never settle amid the excitement of the hunting field, and become a danger to his rider, other riders and himself, while valuable competition horses or racehorses should be hunted with care. There is obviously some risk of injury by a kick, a strain, or any unexpected hazard such as barbed wire or 'blind' ditches.

Your horse's temperament, fitness and stage of education must be taken into account before deciding if the advantages outweigh the risks. It is wise to clip your horse, although he should be fully fit, before taking him hunting, as he is likely to sweat a lot and may lose too much weight, or even catch cold when standing around, steaming hot. If your horse has been in work during the summer he will already be fairly fit, but you should have increased the canter work to get his breathing right.

Plait your horse's mane for hunting, if you want him to look smart, unless you prefer to have some to hold onto, though a breast plate, martingale or simple neck strap will provide that security. Boots or bandages are inadvisable: they become muddy and may slip, or loose grit may work its way underneath causing a sore wound.

Most horses definitely benefit from their varied experiences acquired in the hunting field. They learn to cope with all types of ground in different conditions and at varying speeds, to jump a wide variety of obstacles, to adjust their balance and speed when necessary, and to get used to excitement. Meanwhile the horse develops in strength and power, learns to go forward with enthusiasm and to gallop on. Above all, he learns to look after himself, and therefore his rider, in all situations.

If your horse has yet to experience his first day's hunting, a gentle introduction via cub-hunting is ideal, but you should first get permission from the Master or secretary. There is little action and much standing about, which should help to settle the horse. Remember that at home you want your horse to be obedient to your aids: it should not be any different when hunting, so insist on good manners and avoid hotting him up by galloping

Overleaf To many riders there is no greater pleasure than the challenge of a day's hunting.

Left Hunting in Germany is confined to drag lines with the emphasis on jumping. A small pack of hounds hunt an aniseed trail laid by a runner over a pre-arranged route. Two or three lines will be run during the 'day', which usually lasts about two hours.

about behind the others. A 'follow my leader' attitude is dangerous because, when you want to turn away or jump in a certain place, your horse may not want to leave the others. Ideally, take a sensible 'school master' companion to teach your horse how to behave and to give him a lead over any suitable obstacle you may wish to jump. If your horse has a sensible introduction he will soon learn to relax, but remember he may be keener and stronger to hold out hunting, and you may need a more effective bit in his mouth. Make sure your tack fits well and is in excellent condition: it will be under considerable stress.

What to wear
The hunt staff and officials usually wear red coats to distinguish them and make them easily visible. These are cut fairly long, as protection from the elements in a cold, wintry climate; a lighter material is used in the warmer conditions of temperate countries. The staff wear hunt caps, carry hunting whips with long thongs to help control hounds, but only the huntsman carries a horn to communicate with his hounds and other hunt staff. Tall hunting boots, with tops for staff and gentlemen subscribers, protect the rider's legs in all conditions, on or off the horse. A strong pair of gloves are vital protection against cold or wet, or slippery reins. Spurs are not essential, but a blunt pair is usually worn. A warm hunting shirt or vest with a waistcoat will help fend off the cold and a warm pair of tights under the breeches also help. The hunting stock or

Above *The French huntsmen are renowned for the musical qualities of their hunting. They have many refrains, which they play on their characteristic horns as the day's hunting, usually for wild boar, proceeds.*

tie, correctly tied with a tie pin, completes the picture.

The gentleman subscriber may also wear a red coat if desired, now with a hunt cap, not a top hat. The cut-away version with swallow tails looks very smart and can also be in black, worn with a hunt cap. Most followers wear a strong black coat with a hunt cap or bowler hat (ladies may wear dark blue instead of black). Formerly, only officials, staff, farmers, and subscribers with the Master's permission were allowed to wear a velvet hunt cap, but this rule has changed nowadays. Some people find a hunt whip cumbersome, but it is useful for opening and shutting gates, holding back branches, or if you are helping the hunt staff with hounds. An ordinary whip may be carried instead, particularly if you are riding a young or difficult horse, although it is not considered correct.

If you don't have a dark hunting coat, it is perfectly acceptable to wear a tweed hacking jacket. Before the opening meet (usually the Saturday nearest to November 1 in Britain) clothes are less formal, and 'ratcatcher' is worn (tweed jackets blend well with the countryside) with black or brown boots, hunt cap or bowler.

Comfort and safety are most important, so your clothes should fit well and a neat, clean appearance, as always, meets with favour.

A day with hounds

As healthy exercise, hunting is unbeatable. It is an outlet for the energies of the young and revitalizes the older generation!

Competitive sports such as steeplechasing, point-to-points, hunter trials, cross-country events and recently 'team chasing' all derive from hunting and provide added incentive to breed and train the best: horses with speed, agility, stamina and courage.

If your local hunt is already oversubscribed, or if you prefer not to hunt live quarry, a day with the drag hounds or bloodhounds can provide you with all the enjoyment of galloping and jumping across country. The pack follows a trail laid by a man on foot about one hour previously. For the mounted followers the emphasis is usually on jumping and, although the challenge of the unpredictable is lacking, since the route taken is preplanned and often specially built, many riders are content with two hours' fun. Indeed, many prefer drag hunting and landowners can control the lines to avoid or minimize risk to livestock or crops, which are sometimes damaged by thoughtless hunt followers.

In most cases, anyone who can afford to keep a horse and lives in a hunting area may hunt, so long as they pay the required fee and observe the traditional rules expected of all followers. Ancient rules of dress, manners and language still persist today and, although some seem outdated, they are based on practicality.

The Hunt staff are responsible to the field (mounted hunt followers) for the day's sport. They are also responsible to the landowners and farmers—your hosts, by whose permission hunting takes place—for protecting the countryside you cross. You must make their task easier by obeying basic rules, most of which are common sense.

The huntsman is in charge of the pack of hounds

Far left Hunting in Britain is less formal before November 1 when the season proper begins. It is an ideal time to introduce a young horse to hounds, but is usually by invitation or prior permission and the hunt secretary will give details. It is acceptable to wear 'ratcatcher'—tweed jacket, black or brown boots, bowler hat or hunt cap—providing your appearance is clean and smart.

Above The hunter has to negotiate all types of obstacle, including holes in the ground. The hunting field is a useful proving ground for event horses as it encourages the desire to go forward, without worrying about ditches or other hazards.

Right The huntsman is in charge of the hounds and hunting the pack. He must be allowed to concentrate on his job without having to worry about the conduct of the field.

and the actual hunting. He must be left to concentrate on what each hound is doing when he 'draws a covert' (casts his hounds in a wood, spinney or similar place likely to hold a fox, if this is the quarry). You will hear him call his hounds, each by name. He uses his horn to collect hounds together, to rally them with 'gone away' when the fox has left covert and they are 'laid on' to his line: a different call is used if there is a kill, and again if the animal has 'gone to ground'. Finally, he will blow for 'home'. You will soon recognize what each call on the horn means. The huntsman is assisted by his whippers-in, who keep hounds together and help to maintain discipline in the pack.

The Master is in overall command. He knows the area well, knows the farmers and landowners over whose land you will cross, and most if not all of the followers. He is ultimately responsible for anything that goes wrong, with the duty of preserving and promoting the best elements of hunting, and encouraging the future of the sport. During a day's hunting he is assisted by the Field Master, who is in charge of the followers and who must be obeyed! All field members are expected to be in full control of their horses, to be polite to other followers and in particular to motorists and pedestrians who may be inconvenienced by your activities and can easily become resentful if not treated with due consideration.

How to enjoy your day

Arrive at the meet cool and clean, if possible (allow 6 mph (10 km/h) for hacking on); if boxing to the meet, park considerately and unload a couple of miles away so you can warm up your horse and take the 'edge' off a little. Aim to arrive early and say 'good morning' to the Master and secretary, who will collect your day's subscription money (cap) unless you have already paid the full subscription for the entire season. The surroundings will be very exciting, so be aware of other horses and particularly hounds: turn your horse to face hounds, as a horse which kicks a hound has no place in the hunting field.

If you are intending a day visit, you must first contact the Master or secretary to ask permission to hunt, since the numbers of non-subscribers may be limited to keep the field to a manageable size. You must know in advance how much you will be expected to pay and do so on arrival before you join the rest of the field. If your horse is excited take him to a quiet spot and walk about calmly.

From the meet you will move on to the first 'draw'. Don't chatter loudly when standing beside a covert whilst hounds draw it, and don't make any obtrusive noise or smoke a cigarette. Others will be listening intently for sounds from the hounds, while the smell of smoke may deter a fox from breaking covert. If you happen to see the hunted animal slip away, you may 'holloa' and stand pointing your hat in the direction he has gone to help the huntsman when he arrives with his hounds. If in doubt, do nothing, as the hounds may be on a different line already.

Once on the move it is a good idea to pick out a reliable person and horse to follow as your 'pilot'. Always watch the Field Master, follow where he goes and stay alert. A young or ill-mannered horse

Above *Coyote hunting in California is very much shirt-sleeve order.*

Left *The wheel has turned full circle and many ladies again ride side-saddle to hounds.*

Right *A young horse is introduced to hounds at one of the big hunt establishments in Britain.*

should be kept out of the way at the back. A kicker should wear a red ribbon on his tail, as a warning to other riders, but try to keep him away from trouble. Avoid possible offenders, particularly in gateways or when standing still.

You should observe common-sense rules of courtesy; if you ride on seeds or forbidden land, leave a gate open, overrun hounds or the Master, or are a nuisance in any way, you may be sent home. You must never jump a fence when hounds are not running and if you damage a fence make sure you report it to the secretary. Most manners are a matter of being aware.

Ride as you would at home to make it as easy as possible for your horse, who has a long day ahead and, most important, have consideration for others in gateways, queuing for a jump, when jumping, or on the road. Always take the most economical route without jumping unnecessarily or tiring your horse in heavy or rough ground; the best horseman never appears to hurry, but will always be at the right place at the right time. Watch the hounds; they seldom go straight and if you hinder them—no hunt!

At the end of the day thank the Master, say 'good night' and ride home or back to the box quietly. Your horse will be thirsty, but allow only a short, slightly warmed drink at first. When you get home, rub him down before attending to your own needs. You must examine him for scratches or thorns, particularly on his legs, as to leave them overnight could lead to serious trouble. Make sure he is warm and comfortable and check him again later in case he has broken out in a sweat.

Why hunt?

Many people, even if they are not actively opposed to hunting, consider it an anachronism. Yet more people hunt today than ever; more packs of hounds exist today than at any time in hunting's history. Whether you hunt or not should be a matter of individual conscience and choice, but many who condemn it have not first examined the reasons why hunting takes place. This is not intended to be a treatise on the pros and cons of hunting, but it is a fact that without it many animals would be persecuted ruthlessly by whatever means, sometimes to survive for days with fatal injuries. Hunting is one of the best conservers of wildlife and the countryside; both are protected by the presence of the hunt.

Take the trouble to find out 'why' before you hunt and then, on your fit, well-schooled and well-turned-out horse you will discover that there is much more to it than being the most exhilarating of equestrian sports.

Apart from demanding a high level of horsemanship, hunting will serve to improve your riding technique as the experiences encountered will encourage you to make quick decisions while going flat out.

WESTERN RIDING

The original western riders were workers, not skilled horsemen. They were the cowboys, who recognized that though they might understand the need to move cattle, only the horse could get the job done. A rider who tried to dominate or direct his horse fell behind, lost his cow, and frustrated his mount. So the basis of western riding is that the horse works free rather than under the control of a rider. The rider requests; the horse responds.

Western horses must be trained to understand the communication system so that they can understand the request. The riders recognize and understand the anatomy of the horse. They know which aids will elicit a particular response from nerves and muscles. They make the decisions, but the horse carries out the task.

What is a western horse?

Almost any of the light breeds will make a good general-purpose western horse. American Quarter Horses, however, are considered to have an edge. They are said to have more 'cow' than other breeds, and as they have had more cattle working experience as a breed, the notion is probably true.

The best conformation for the western horse is that which promotes quickness and agility. Horses 14.2hh to 15.2hh are the best size for working western horses. They need good bone structure with knees and hocks low to the ground, and bulky, relatively heavy muscle is best for speed when there is little need for endurance.

Tack and its origins

The stock saddle is comfortable, functional and durable. It had to be for it was ridden for hours and hours on end, protected its rider from sweat and chafe, was the anchor which held many a steer, was the carry-all for slickers, ropes and hobbles, and, at the day's end, often served as the pillow for the tired cowboy.

Specialization brought many subtle changes to the stock saddle. The roper must stand in his stirrups as he throws his loop, and therefore the roping saddle has little swing in the fenders. The pommel of the cutting saddle is high, square and almost sweeps back to the rider's thighs, holding him in place. Today's western game saddles are cut smaller and lighter for the barrel racer and pole bender.

The 'equitation' seat is heavily padded between the swell and the cantle. This padding locks the rider down in the seat, giving him the feeling of being securely set in the 'proper' position.

While all such creative innovations have their advantages, they are restrictive. The general purpose or ranch saddle, however, offers a pleasant ride without restrictions imposed by specialization.

The saddle should, of course, properly fit both horse and rider. The average rider will find a 15- or 16-inch (38–40 cm) seat comfortable. A smooth swell on the saddle, 12–13 inches (30–33 cm) in width, is very practical if a variety of exercises are to be performed. All other factors being equal, a saddle weighing 30–35 lb (13–16 kg) makes the best working saddle. A second cinch on the western saddle serves the purpose of keeping the back of the saddle from rising when a roper dallies.

Breast collars are a good place to stick silver, but can impair the breathing of a fast-working western horse. It is, however, necessary to use a breast collar on a horse with faulty conformation.

Young western horses are trained in a snaffle bit. Finished western horses work in a mild curb, generally a version of the grazer, which has a swept-back shank. Half-breed and spade bits are used on the finished horses trained to the California style of head and neck carriage.

Western work

The work of the western horse is ranch work. The days are long, hard, and often hot and dry or wet and cold. And the work doesn't stop just because the sun goes down; many a ranch horse has to work all night, slowly walking around a herd of cattle while a tired cowboy sings a lullaby.

The western horse works in three gaits; the walk, the jog and the lope. The western walk is not brisk, but covers ground and is steady in its pace. The

hind foot should well over-stride the front hoof print.

The jog is not a natural gait, but must be mastered. It is a two-beat diagonal gait, much slower than the trot. Its pace is steady, and the good western horse should be able to jog for hours. This is the gait used most when herding cattle; it is slow because cattle are moved slowly.

The lope is a slow three-beat gait which covers ground, is easy for the cowboy to ride, and does not have the tendency of making other animals nervous or want to run.

The western horse must also be good at backing. He holds a calf or a steer by backing and keeping the rope taut, and likewise often pulls a hay bale or logs. And the horse gets out of difficult spots not by wheeling and running, but by backing away.

Variety in competition

Western horses compete in a variety of events. There are, of course, the extremely popular pleasure classes in which the horse works all three gaits in both directions of the ring. There are trail classes in which the horse must demonstrate his ability to position the rider to perform any number of ranch or trail chores, such as opening and closing gates, and negotiating obstacles such as logs and streams.

There is western riding in which the horse's handiness is tested as he completes a pattern calling for eight flying changes of lead, a solid stop, and a good back. Reining classes demonstrate speed, agility and patience as the horses are asked for flying changes of lead, spectacular slide stops

and fast spins, all punctuated by periods of rest on a loose rein.

Barrel racing and pole bending are typical games for western horses. Again, speed is required along with a disposition which allows a horse to walk quietly away after a contest.

A working cow horse is the epitome of the western horse. He must demonstrate almost all the manoeuvres unique to the western horse. He must show his calmness and his patience by moving quietly and slowly to a cow, never anticipating his rider's decisions. Once asked to hold the cow, the horse should work it in a small area, without rider assistance. Then, when asked, the horse should run the cow down the arena, turning it back by cutting in front of it and performing a rollback at high speed. The final test of the horse's agility and quickness comes when he is asked to move the cow to the centre of the arena and turn it 360° to the left and then to the right.

The ultimate western horse is the cutting horse. The cutting horse moves into a herd of cattle without disturbing them. When shown which cow to work, the horse locks his attention on that particular animal and, with no further assistance from the rider, removes and holds the cow from the herd, blocking any attempt to return. Any assistance by the rider is a fault and is penalized. The horse must work in total freedom.

Riding Texas style

There are two styles of western horse, and two styles of western riding. Each has its following. The Texas style was developed by the American

Overleaf The job of the Western horse is ranch work.

Above left The Texas style has always been the most popular amongst Western enthusiasts, and was developed through the needs of the cowboy who knew little about his horse except that life was impossible without it.

Above The reins, held in either hand, are taken in the palm, the fingers are closed and the hand turned over so that the wrist is straight and the thumb up.

cowboy, who knew little about his horse except that life without him was nearly impossible. So the cowboy went everywhere and did everything with his horse; it was a full partnership. When the cowboy could do it better, he did it; when the horse could do it better, the cowboy turned him loose.

The Texas cow pony was a compact horse with plenty of muscle. The basis for one style of American Quarter Horse, the cow pony was a Thoroughbred crossed with a small Spanish mare. They were cat-quick, knew cattle, and could stay fat on range grass.

When the cowboy wanted to get some work done, he asked his horse to do it, then he pitched his reins away, and the horse went to it. This has always been the most popular style of western riding in the United States, probably because the Texas cowboy and his horse drove cattle over great distances, and demonstrated their skills to whoever might watch.

The saddles used by these cowboys were not fancy. A good, safe, comfortable rig was what was needed. The cowboy had little time to care for his tack, and he certainly had nowhere to keep it away from the elements.

This style calls for the use of light, narrow, split reins. The bit is usually, solid-jawed, low port curb with a swept-back grazer shank.

When riding Texas style, the reins are held in either hand. The reining hand is placed palm up under the reins, and then the fingers are closed around the reins. The reins are not grasped tightly, but are held firm between the thumb and index finger. The reining hand should then be rolled so the wrist is straight and the thumb is up. The loose ends of the split reins are dropped on the same side of the horse's neck as the rider's reining hand. If the rider is reining with the left hand, the loose ends of the reins are on the left side of the horse's neck. He may not touch the reins with the right hand, which is carried either at his waist or straight down at his side. If a dismount is called for, a Texas-style rider will leave his horse ground-tied by dropping one of the reins on the ground.

The Texas-style horse travels in a very natural way. He carries his neck straight out from the withers, flexing at the poll, but not bringing his nose into a full vertical position. The rein hangs in a loop from rider's hand to the bit shank.

The horse responds to the indirect rein. If the rider wishes the horse to move to the left, indirect rein pressure is applied. The horse moves gently and easily to the left. If more speed is required, or the turn is to be made tighter, then the rider asks the horse for the additional work by applying leg and weight shift cues.

The key to training the western horse to the Texas style is in never applying a steady pressure to the bit through the reins. The horse must always be pushed to the bit by the rider's legs. He can never be allowed to reach the bit unless it is to discontinue forward movement.

To position the horse's head, the Texas-style rider will tug and release, tug and release on the reins while applying leg pressure to push the horse forward. The tug and release cue will soon have the horse dropping his head and neck into the proper position if the rider pushes the horse forward with

Above left The California style was developed by the immigrant Spanish ranch owners who had a gentleman's background in equitation. The horse schooled in the California tradition moves entirely in self-carriage, having no contact with the rider's hands.

Above The reins can be held in either hand. Taken in the left hand they are held principally between the thumb, which is upper-most, and the index finger. The free hand is used to grasp the romal, the integral whip, at least 40 cm (16 in) from the point of attachment to the rein.

leg aids. The finished Texas-style horse will drop his head into position at the jiggle of the reins.

The Texas-style horse moves forward on a loose rein, only occasionally being reminded to keep his head position by a slight jiggle of the reins. If the horse reaches the bit and the reins become taut, the horse will stop.

Riding California Style

The California style of western riding was developed by the Spanish Vaquero, who had a gentleman's background in horsemanship. The Vaquero wanted more than just a working horse, he wanted a flashy, prancing, spirited horse, and he was willing to spend the time to make one.

The Vaquero preferred a horse with Thoroughbred conformation, so the California-style horse tends to be tall and light with a graceful neck.

The California-style horse begins his training in the hackamore, the Spanish training tool. The first bosal (rawhide braided noseband) is large and heavy, and the horse soon learns he is under the rider's control. As the horse progresses in his training, smaller, lighter bosals are used. Finally, when training is nearly finished, the horse is tacked with a half-breed or spade bit and a pencil bosal. The horse carries the bit, but is ridden on the bosal. The gradual changeover to bit may take a year or more. When the Vaquero is sure the horse's mouth is perfect, the horse is considered finished. When the horse is switched to the bit, his mouth is said to be 'made'. It is sensitive to the slightest movement of the bit, which is never used as a device for punishment.

While the horse is in the hackamore, he is taught to carry his head in a vertical position with the neck arched and the third vertebra higher than the poll. To accomplish this the rider must push the horse forward with his legs while applying a direct rein of opposition. The horse should arch his neck, round his back, and move his hindquarters under his body. If correctly done, the horse's centre of balance will be moved back under the rider and the horse will be in a collected position. The California-style horse moves in a much more collected position than does the Texas-style horse.

When the California-style horse is finished, there is no need for the direct rein of opposition. The heavy bit will hang in the horse's mouth properly from the headstall if the horse tucks his head to the vertical and assumes a collected position. The reins are loose and the properly held bit is the barrier behind which the horse remains.

The key to training the western horse to the California style is in overflexing him. He must be taught to stay behind the bit, no matter what the bit position. If the horse will overflex, he can then be pushed forward to the bit on command. But if he disregards the bit and moves into it, collection on a loose rein is impossible.

Far left Slide boots protect the hind fetlocks and heels of the horse as it gets its feet right under its body for a slide stop.

Left The lightweight curb used in Texan style riding has shanks that are swept back to enable the horse to eat in comfort when wearing its bridle.

Left The roll back is a half-turn, pivoting on the inside hind leg. At whatever pace the horse starts the movement it must continue without hesitation on completion.

Below The spin, a full 360° turn, is a complete turn on the haunches and is practised slowly at first by making a series of quarter turns, or 'off sets'.

The California-style rider often uses a tooled saddle with plenty of silver. The headstall is usually well decorated with silver, and the bit is always silver with large, intricately designed shanks. The reins are braided leather and are closed. The romal, a long leather braid which can be used as a whip, is attached to the closed reins.

The California-style rider grasps the reins in either hand. If the reins are taken in the left hand, the hand is placed palm down on top of the reins and the fingers are curled around the reins, which are held principally by the thumb and index finger. The back of the hand is then rolled towards the horse's head so the rider's wrist is straight and the thumb is up.

The romal is grasped in the right hand, at least 16 inches (40 cm) from the point of attachment to the reins. The rider carries his right hand on his thigh.

If asked to dismount, the rider lays the reins on the horse's neck, and places the romal across the saddle behind the saddle horn from right to left. A California-style rider never ground-ties his horse; instead he uses hobbles, which he carries on his saddle at the pommel just in front of the fender.

Slide Stops

The performance skills unique to the western horse stem from his work with cattle.

The slide stop is a dramatic extension of the hard, fast stop required of the horse used by the cowboy when roping. It is hard on the horse, and should not be practised too often. Good, clean stops should be expected from the western horse, but too many slides will soon make him sour.

The slide stop is not accomplished by pulling hard on the reins. It is the result of the horse moving with speed so his momentum slides him along the ground when he gets his hind feet well under his body. The horse's head and neck should be elevated, and his front feet should continue to 'walk' as his hind feet slide. The horse should round his spine as he tucks his tail towards the ground.

The slide stop should be practised first from the walk. The horse should stop completely each time the rider gives the cues. If the horse takes a step forward after the cues are given, the rider is failing to school him correctly. The horse must stop immediately and completely.

It is a matter of practising form and style, not speed and distance. One or two slide stops can be attempted each training session, but only after several months of perfecting the slow but correct western stop. Once the horse is stopping well, he should be shod with sliding plates, and he should always wear skid-boots when working.

Below The slide stop is a dramatic extension of the hard, fast stop required of the cow pony when roping, but should be practised in moderation. Form and style should be practised, not speed and distance.

Above Brushing or splint boots are worn to protect the horse from interference injuries when practising the rough Western performance skills.

Offsets and rollbacks

Offsets and rollbacks are used by the western horse when he is cutting cattle. They lead to the spectacular western spin.

The rollback is a 180° turn over the horse's inside hock. If the horse is going to rollback to the right, the rider must first give him the cues to stop. As the horse is preparing for the stop, the rider must use right leg pressure to ask him to advance his right rear foot. As the horse stops, the right rear foot should be slightly more forward than the left rear, and he should be in position to snap around over his right rear hock. The rider still has the reins lifted from the stop cue, and now gives the horse the indirect rein cue to turn right, but without forward movement. The bit position, held steady, prevents the horse from going forward. The rider turns his upper body and looks back in the direction he wishes to travel. In turning, the rider's right leg will drop back, holding the horse's right rear foot in position. The rider's left leg will move forward and should now apply pressure at the girth. This pressure pushes the horse's forehand round. The horse should cross the left foreleg over the right as he makes the 180° turn.

If the horse begins the rollback at the jog, he immediately resumes the jog when he has reversed direction. After the rollback is complete he should immediately pick up the previous gait.

Offsets are 90° or quarter-turns performed in the same way as a rollback, only from a standstill. To do a quarter-turn to the left, the rider asks the horse to move the left hind foot forward slightly. Then the rider gives the horse an indirect right rein cue to get him to move the forehand to the left. Speed is added to the movement if the rider requests it by applying right leg pressure. The reins are lifted as a barrier to forward movement.

The rider should turn the upper body in the direction of the movement, and must be careful to discontinue all cues quickly so the horse does not go beyond the 90° position.

In a reining performance, a western horse might be asked to do a quarter-turn to the left, followed by a half-turn to the right, and then another quarter-turn to the left. There should only be a slight hesitation between each movement.

As with all other exercises, spins are practised slowly, with speed added only after the horse knows exactly what is expected of him. A rider cannot force a good performance from a western horse. The horse must want to perform as requested, and he must be free to perform.

If the horse is going to spin to the right, he must place his right hind foot slightly forward. The rider cues the horse to do this by applying right leg pressure. Then the rider gives indirect left rein pressure while lifting the reins slightly to prevent the horse from moving forward. The horse begins to move to the right, away from the rein pressure.

The rider applies left leg pressure at the girth and the horse moves the left foreleg across in front of the right foreleg. The horse then moves the right foreleg, and finally pushes off hard with the left hindleg. The sequence is repeated again and again as the horse completes 360° turns.

The western horse can be taught the spin by linking a series of offsets together, or by asking for a slow 360° turn. In a western spin it does not matter how many strides the horse takes to complete the turn. Speed is desired as long as the horse remains low while spinning. A horse which jumps around is not performing well.

Once the horse can execute a good slow spin, speed can be added by working the horse in a circle at the jog, then making the circle smaller and smaller until the horse is performing a spin. For more speed the horse can be started at a lope.

The rider can get the horse to spin with speed by continually bumping him at the girth with the

Below The California horse is required to work in an over-flexed position. Here the draw reins are used to obtain the required head carriage with the third vertebra higher than the poll.

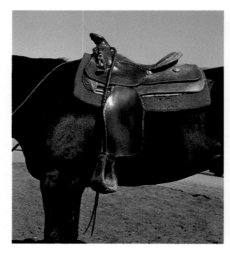

outside leg and applying a consistent indirect rein pressure. The horse should not stop spinning until both leg and rein pressure are discontinued.

The western horse should back easily and rapidly. To back the horse, the rider must first get the horse's weight shifted to the hindquarters. This is done by shortening the reins, not pulling back. The reins should be shortened until the horse shifts his weight; then they should be held steady.

Once the horse has shifted his weight to the hindquarters, all the rider need do is squeeze with both legs and lift his body slightly out of the saddle so the horse can round his back.

The sidepass is the only exercise in which the rider actually positions himself behind the horse's movement. If the rider wishes to sidepass to the right, he must shift his body weight to the left, apply

left leg pressure, and indirect left rein pressure. He must lift the reins slightly to prevent forward movement, but also must be careful not to shorten the reins so much that the horse steps backwards.

When performed correctly, the horse will cross both the front and rear left legs across in front of the right legs when moving to the right. Neither the forehand nor the hindquarters will lead. The horse should move sideways towards or away from an obstacle, such as a gate.

The western horse is not properly trained if the trainer attempts to force him or exert control over his every movement. He is not properly ridden if the rider fails to give him freedom. But free of domination, and with the correct degree of handling, he can respond to his rider's every request and give a brilliant performance.

Above *A working cow horse is the epitome of the Western horse. With the cow roped, the horse must make a hard, fast stop or slide to bring the animal to hand.*

Left *The Californian style demands elaborate saddlery, like this show bridle.*

Far left *The functional working saddle; the safe, comfortable rig required by the Texan cowboy.*

10

LONG DISTANCE RIDING

Long-distance riding must have the oldest origins of all equestrian sports. The first man (or was it a woman?) who leapt onto the back of *equus* must have had a pretty terrifying ride for many miles at high speed. The fact that an animal could provide transport which went further and faster than a human on foot ensured its survival.

Over the centuries many tales have been told of incredibly long army marches and migration of tribes, and many a doctor must have had to thank his horse for getting him to aid a sick patient or infant's arrival in time. How history might have been altered had it not been for the courageous horse carrying vital messages over miles of hostile mountain territory, battlefields or parched deserts; not to mention the famous Tschiffley and his 10,000-mile ride across the Americas with his faithful horses Mancha and Gato.

Today's sport should carry on these traditions, and the aim of every competition should be to keep alive the good qualities of the past—to accomplish the task without undue stress, and always in a condition 'fit to continue' should the necessity arise.

Basically, good conformation is a necessity because, as in any type of mechanism, animal or human, the horse cannot work at maximum efficiency if there is a weak link in the chain; the top-class steeplechaser seldom has a narrow chest and weak hocks.

In America the Arab has pride of place and features in the top placings of most of the national endurance rides, but this may be due partly to the high percentage of Arabs competing, and certainly they seem to cope well with very high temperatures and humid climate. Additionally, certain Arab breeders concentrate on breeding for endurance performance horses which fetch very high prices, something which could be emulated more widely in other countries. In Britain it seems to be the type of agile small horse, about 15hh–15.2hh, that is consistently successful. These are mostly native ponies crossed with Thoroughbred or Arab, or pure Thoroughbred and Arab, and now a sprink-

ling of Morgans, Appaloosas, Quarter-Horses, Trakehners and Haflingers are making their mark.

With these thoughts in mind, a very critical look should be taken at the prospective long-distance horse. Particular attention should be paid to a good strong framework—plenty of room for heart and lungs which will be called upon to function to their maximum capacity, well-proportioned limbs with strong, flat joints and short cannon bones for least amount of stress on those vital tendons. Pasterns should be just the right angle, not too upright (concussion), nor too sloping (weakness), ending in hooves well formed, hard and all of equal size. The hind end is the engine and therefore must be exceptional if it is to function to propel the horse and rider for 50–100 miles (80–160 km) efficiently.

Last, but by no means least, the head should be studied as it can tell a lot about the character of the horse. Large eyes denote generosity and equable temperament; small ears, intelligence and vigour; wide forehead, brains; large nostrils and width between the jaw bones enable vast amounts of air to be inhaled to fuel the engine; a good set of teeth with level bearing surfaces enable good mastication of essential nutrients in the (very expensive) diet. For the long-distance horse, remember the four important 'Cs'—Constitution, Comfort, Character, Condition.

The competition structure

In America and Australia, long-distance riding has been a thriving and popular sport for over thirty years. The competitions are organized under various associations with strict rules to safeguard the welfare of the horse. As standards rise and veterinary knowledge is increased, so the distances and speeds have been increased. Both countries run competitions nationwide for all categories of riders, including the top prestige 100 miles (160 km) in a day rides of the Tevis in California and the Quilty in Australia. South Africa also has an active long-distance group with an annual 100-mile (160 km) ride.

Britain entered the field of long-distance compe-

tition in the late 1940s when several organized rides met with good response both from competitors and sponsors. In the recent years, the sport has spread to France, Germany, Italy, Switzerland, Spain, Portugal and Belgium. Each country has a long-distance association providing rides of all categories, including a top prestige ride, while most of their rules are based on those of the American pattern. The European Long Distance Riding Association (ELDRA) sanctions rides in each country and runs an annual championship based on points gained from two competitions in the home country and one abroad. The FEI is currently considering proposed international long-distance ride rules in the hope that it will become an official competition.

What is involved?

Long-distance riding can involve every type of horse or pony with riders of varying ability.

Pleasure and training (non-competitive) rides of 15–25 miles (24–40 km) are admirable for first-

Overleaf The long distance horse should have comfort, constitution, character and condition.

Above Handsome is as handsome does: the small horse with strength, agility, a calm yet generous nature, soundness and an iron constitution is consistently successful.

Left The popularity of long distance riding, which had its foundations in the States and Australia, has now spread through Britain to France, Italy, Germany, Switzerland, Spain, Portugal and Belgium. Each country organises its own rides of all categories over all types of terrain.

timers to see if they like it, find out how their horses shape up and meet other riders. Very experienced riders sometimes use these rides for exercise work and they are normally only too pleased to pass on information and give helpful advice.

Most long-distance riding associations run a programme of rides under their own rules with distances of 25–100 miles (40–160 km), the longer mileages sometimes being divided into a two- or three-day ride. The judging is basically under two categories.

Competitive rides are ridden within a time bracket over a set mileage on a marked course and judged at the final vetting on condition, pulse/respiration rates back nearly to pre-ride, evidence of tiredness, sweating and so on.

Endurance rides (against the clock), again on a marked course of 50 miles (80 km) minimum; longer mileages are sometimes divided into two days. It has been found, contrary to what most people would expect, that the shorter distances of 50–75 miles (80–120 km) are the most dangerous; it is the speed not the distance that puts the horse in

Above Even the fittest endurance riders and horses need to take a break for refreshments along the ride route. This is where the back-up helper becomes indispensible, for it is he or she who will be on hand at the pre-planned destination with the much-needed provisions.

jeopardy. The speeds achieved today on a really fit horse with a competitor out to win can be much too fast for safety, especially over rolling downland. With 100 miles (160 km) plus before them, competitors ride more cautiously and at slower speeds, and the horses are able to cope very well within their own capacities.

Training and preparation
The training and preparation of the long-distance horse need as much knowledge, perseverance and dedication as preparing a horse for jumping, dressage or eventing. Not only must the horse be superbly fit but he must also be well balanced, move well in all paces and be immediately responsive to the rider's aids and wishes. He must be able to perform at maximum speed and efficiency over rough, stony, slippery ground, soft and hard going, with steep gradients up and down hills. Natural hazards have to be negotiated; fording deep rivers, passing through busy farmyards. At all times he should be 'in tune' with the rider and prepared to go first or last or away from others. He must not get too excited at the beginning of the ride-and wear out both himself and his rider in the first 10 miles (16 km).

School work should include lengthening and shortening the stride, at your request and on his own initiative, maintaining balance through transitions, while the rider should recognize the change of diagonals in trot and leads in canter. Lateral movements—or moving away from one leg—may cut seconds off the time if gates can be opened and closed quickly and efficiently.

The whole process of preparation must be gradual. The first period of work should be on the same lines as getting the hunter fit (see Stable Management). Then it is the day-after-day build-up of preplanned work; long, slow days with shorter, faster workouts (with point-to-pointers if possible). Any lost days, for whatever reason, can never be made up in the countdown programmed for the competition season. Every competitor has to fit this programme of exercise, feeds, grooming and general stable management into their own lives of family and work commitments.

An average work programme should read something like this:
Daily exercise—½ hour lungeing with 1½ hours hacking (varied routes).
One day—a longer 3–4-hour ride.
One day—approximately 3 hours with a half-way break.
One day—speed work, i.e. working up to 1½-mile (2.5 km) gallops.
One rest day.

Daily exercise can be made interesting for both horse and rider by incorporating different types of terrain (even if you have to haul to it) and the use of school movements down quiet country lanes, such as lateral work, serpentines, transitions at predetermined posts and, of course, the all-important, time-consuming gate opening and shutting.

During all this training it is necessary and of great interest to note the amount of sweating, the effect of humidity, the breathing in relation to the horse's stride and pace, and of course to monitor the pulse and respiration rates at the end of exercise, and again twenty or thirty minutes afterwards. It is this last reading, when compared with the reading of the horse at rest, that will give you guidelines on his true fitness. Various other forms of activity such as hunter trials, eventing, jumping, hunting and driving are available to the fit and versatile horse.

Swimming can be a very useful form of exercise in the fitness programme, especially when the ground is wet and heavy, or exceptionally hard due to drought or frost. It is inadvisable to try to use any pond or river, as the bottoms can be muddy and choked with weed, the current too swift, and the access and exit difficult for the horse to negotiate. If you are near enough to the sea and a good sandy beach you are very fortunate, but if not, there are now specially designed equine swimming pools, particularly in racehorse training areas. These are excellent as they are administered by professionals who know exactly for how long your horse can swim per session, and they will build up an exercise programme over several weeks. They give a complete service while the horse is in their care, putting him into protective boots before the swim, providing two attendants to lead him in and dry him off afterwards under infra-red lamps. The horse is immensely refreshed; heart, lungs and muscles have had maximum exercise while tired tendons and legs have had a rest, and it is of course excellent for horses laid-off normal work due to minor injury.

Shoeing The inevitable vast mileage involved in getting the horse fit and the subsequent competitions will mean many visits to the farrier. He will need to be very skilled in his trade and sympathetic towards your requirements. The long-distance horse may need corrective shoeing if shoes are worn unevenly, or to alter faulty action. Obviously the horse that travels well and is light on shoes has a distinct advantage over one that wears out shoes for a pastime or, even worse, keeps casting shoes. A good farrier will also give advice on the quality of the hoof wall, which may need help through a feed supplement or hoof dressing. Most people have to do a certain amount of roadwork, necessitating the use of road studs, but these should be fitted to the hind shoes only, not in front as they would cause too much concussion. When all else fails and the horse is still wearing out a small section of shoe long before the rest, the addition of titanium or borium welded on before setting the shoe can be helpful. With the horse that has thin soles or suffers from concussion, pads can be fitted between hoof

Above There is great companionship in long distance riding, but there can also be periods of great solitude. Psychologically the rider must be prepared for this and the horse too must be a willing performer on its own, able to lead or follow without becoming agitated.

Above A well-shaped and well-shod hoof. The frog is distinct and healthy and the balanced design of the shoe indicates good conformation in the horse. 'No foot no 'oss' goes the saying, and considering the vast mileages the long distance horse has to cover this is particularly relevant.

and shoe and these can be very beneficial, particularly where the course is over an area renowned for stones and flints. Pads are not advisable on soft going, deep heather or large boulders as the shoes cannot be fitted so securely and there is a far higher risk of losing one.

Feeding It is a good plan to find a staple maintenance diet for long-distance horses, and then work on the basis that a high-protein diet is necessary for day-to-day high performance, such as the training programme to repair and rebuild muscle fibres. About a week before a big ride start to run down the protein and boost the carbohydrates (energy feeds). Work your feeds into your daily routine— four feeds are better than three—with water always available, but take note of consumption. Supplements should be fed to the high-performance horse, but the best one to suit the individual horse is a matter of trial and error—take advice from your veterinary surgeon. Salt should be given at a rate of approximately one dessertspoonful divided between, and well mixed into, the feeds each day. Give rock salt if available but not the mineral blocks

if another supplement is being fed. Glucose in chilled water given after a hot or long exercise will help to stop post-ride sweating. If half an hour's good clean grazing is possible, this will provide natural minerals and trace elements in the grasses and herbs, as well as beneficial relaxation in a roll and a frolic.

Before competing in a ride, the horse should be fed a late supper, only a token breakfast, and hay should be removed early on the morning. During the ride there is usually some grass available at the halts, and on a long ride succulent hedge pickings can be offered while still travelling as well as small feeds at the compulsory halts. After a competition, a horse who is fit and well will be ready for his feed as soon as he is cooled out, whereas a horse under stress will take some time to regain his appetite.

Saddlery If the horse were to be given priority, the American western or Australian stock saddle would be the most popular, as these have evolved over many years and are strongly constructed with wide bearing surface and deep channel. The girth (cinch) attachment is rigged, giving even distribution of weight, and anchors the saddle to avoid any swinging, rolling or tipping—the main causes of sore backs. Also they fit *all* horses, *but*—and it's a big one—there is a weight problem as most of these saddles weigh over 30 lb (13 kg). American saddlers have come up with a lightweight edition for endurance riding, but this does not seem to have caught on in Europe so far. Australia has an excellent stock saddle with all the advantages of the western without being so heavy. Stirrup leathers and stirrups should of course be of the best quality and girths must be chosen with care and kept scrupulously clean. Horses react differently to various types of leather, nylon, cotton, mohair—again, trial and error will find which suits your horse best. Numnahs are another problem, and most competitors have a special preference, but foam and nylon are not advisable as they tend to 'draw' a hot back.

The choice of bitting or bitless bridle is again an individual one. Bitless bridles have the advantages that the horse can eat and drink better *en route*, you do not risk veterinary penalties for sore mouths and the horse can be tied up safely for halts or an emergency on course. There are now some very good nylon and cotton web bridles coming on the market which should be a great asset to those who have to exercise every day in every sort of weather condition.

Preparing the rider

If the rider is to enjoy the competitions, fitness is essential. It is also essential to get your horse fit and do the relative grooming, mucking out and so on, but you yourself need daily exercises which help to keep muscles supple and in tone. If you're above average weight you should try to lose the extra pounds in the interest of what the horse has to carry, as well as your own health, and your diet can be based on the same principles as for the horse—high-protein during the preparation and then a high-carbohydrate just before the competition to boost the energy levels! Fluid intake should be restricted twelve hours before a ride—there's no time to seek a handy bush *en route*, but facilities are usually provided at the mandatory halts. During the ride your back-up helper should provide you with whatever food and drink you find most suitable, and salt tablets may be necessary in very hot weather.

Below *Schooling should include thorough practise in opening and shutting gates. A schooled horse and rider should negotiate a gate in no more than 30 seconds. Fumbling with a gate can waste much valuable time, as well as cause annoyance to other competitors.*

Right *The rider's clothes are a matter of choice and comfort, but all riders should consider fitness a priority if they are to enjoy their competititons to the full.*

Clothing must be carefully chosen and well tried beforehand to cope with hot, cold and wet conditions. Hard hats are obligatory on all rides. Check the rules of the ride organizer for any clothing restrictions.

Last, but by no means least, how is your mental attitude? The physical and mental stress built up over a long period of time throughout the competition need to be counteracted by great stamina and determination.

The back-up team/helper In all competitions of 40 miles (65 km) plus it is necessary to have help: the horse may require water, if there is no natural source *en route*, and other items may be required such as replacement tack if broken, or spare shoes if lost. The helper must be briefed adequately before the ride and provided with a local large-scale map, the ride route and description, together with your ride time worked out. The water should be readily available for the horse (to drink and cool off in hot weather) *as you meet up*. These helpers spend hours, sometimes in pouring rain, freezing fog or blazing heat waiting for their competitor, for little glory or reward, and should be given due consideration and thanks.

Ride strategy Competitive rides require considerable pre-ride planning. The exact mileage must be known, where the check points and veterinary inspections *en route* are to be held, whether or not time is allowed for these in the max/min time bracket, and where there are suitable places for the back-up helper to meet horse and rider with necessary equipment and water. This will entail many hours of map study to ascertain the type of terrain, accessibility by roads and where natural water is to be found. The rider should carry the ride map and description in case of emergencies, such as vital markers removed by vandals or adverse weather conditions. A small reference card can be attached by sticky tape onto the arm next to the watch, and should contain details of check points, mileage and the time, helping you to keep a check that you are maintaining speed while riding along. You should allow enough time to cover the last few miles slowly in order to bring the horse in cool and relaxed to achieve vetting parameters without loss of marks. **Endurance rides** are ridden on tactics and assessment of the other riders. It is almost as important to know how they operate as it is to know how fast and for how long you can ride your own horse. The object is to complete the ride in the fastest time, but it's no use doing this if the horse runs out of energy or finishes lame.

In these types of rides it is imperative to *know your horse*, but even more so for the American **Vet Gate Rides**. Basically, this is a ride against the clock, but, instead of the usual mandatory half-hour halts every 20–25 miles (32–40 km) along the route, there are veterinary checks at pre-specified places approximately 12–15 miles (19–24 km) apart. It is up to the competitor to know exactly at what speed the horse performs at his best, in conjunction with the most economic pulse/respiration rates, so that at the vet gates you can pass through with the minimum rest period to conform to the parameter rates laid down for the ride (usually P64—R32). This is a form of 'race' but very much with the brakes on. The name vet gate is what it implies—no horse may pass through the veterinary pen and proceed on the course unless it has been inspected and has passed the standard required. This puts the onus on the competitor to present his horse in a fit state to continue, for if he does not meet the veterinary requirements immediately there is a hold-up before he can be re-vetted, thereby losing vital minutes. This can never be made up in the next lap without overstressing the horse, who will then require even longer to rest in order to meet the P/R standards. Truly, 'the fittest shall be first' in this competition.

Long-distance riding probably has more to offer to the general riding public than any of the other disciplines, which are becoming increasingly 'professional'. In distance riding the welfare of the horse is paramount. It can appeal to every type of rider from the person who just likes to hack quietly on a fit, well-cared-for horse, to competitors from other disciplines (for example, to test the fitness of the eventer, relieve boredom/tension in the dressage horse with some natural free extension, and to help rehabilitate stale show jumpers) and to the competitor and horse who reach out to accept the challenge of 'further and faster'.

11

SHOWING

To those who have never been involved in showing it may seem extraordinary that people are prepared to put in many hours of preparation and incur considerable expense, not to mention standing often for hours on end in inclement weather, just to try for that all-important rosette. But to the devotees it is an enormously interesting part of the whole equestrian scene.

Showing is not undertaken for financial gain, as the prize money is often small. Shows, however, are valuable market places for breeders and professionals to advertise their wares, and for the amateur owner/rider they offer the challenge of turning out a horse that is the best on the day.

It is divided into two main categories, showing in-hand and showing under saddle. All breeds, ages and types of horse are catered for from Shetlands to Shires; part of the fascination is that different judges will prefer different horses, so you may be up one day and down the next.

What is a show horse?
So if you want to show your horse or pony, what is the best way of going about it, and what makes a horse a show horse? The main point to bear in mind is that a show animal is one that is as nearly as possible the perfect type for the class in which it is entered; for example, a show hunter should be the classic type of horse that you would choose for a season's hunting, with good conformation and movement and an ability to gallop with untiring ease. So, is your horse a hunter, a hack, a riding horse, or have you a pure-bred Connemara or part-bred Arab? Which of the show classes, or types, or breeds do you like the best? Whatever you choose, there will be a class for it. Each of the breeds or types has its own society which lays down rules and regulations, and you should register with the appropriate one. If in doubt, never be afraid to seek professional advice. Your horse may be good enough to do well at the smaller shows, but not of the standard to shine at the major events. Be realistic; it is better not to waste time and money at large shows if you have little or no chance of being

placed, when you can get a lot of pleasure and satisfaction winning at the smaller shows. Good show horses should be unblemished but in working hunter classes splints and scars are acceptable, providing they do not interfere with the action and are not unsightly, and a 'clean' horse will usually come out on top.

If you really want to go to the top, the business of showing becomes very expensive, and these days most people like to own a horse that is capable of being an all-rounder; one that could, say, win a working hunter class and do well in novice dressage, one-day events and some show-jumping.

So how do you decide if the horse you have is good enough to show, even locally?

Good conformation Whatever breed or type your horse is, it must be well put together to stand up to the scrutiny of the judges. A show horse must have presence, a good front, with the head, neck and shoulder correctly set-on. It should have a middle which is not too long or shallow, with a well-sprung rib cage and plenty of heart and lung room, and strong well-developed hindquarters with a well set-on tail. It should have very good limbs and feet and the whole picture it presents should be one of symmetry and balance.

Faults to be avoided are short thick necks, straight shoulders, long spindly forelegs and bent weak hindlegs set away from the horse.

Good movement The horse should move straight and freely, both in-hand and under saddle. He should have a good active walk with definite strides; a rhythmic flowing trot, moving well off his hind-end, and an comfortable canter and gallop.

Bear in mind that in most adult classes the judge may well ride the horse, so his paces—and manners—must be good to give a good ride. That the horse should be well schooled is vitally important, and this is dealt with later.

True to type It is a waste of time to put your horse in a class for which it is not suitable—e.g. a hack would be totally out of place in a hunter class, so having selected your horse and what category it falls into, don't put it out of its element.

Producing the horse for the ring

It is as much of an art to produce a horse for the show-ring as it is for a racehorse trainer to produce a horse fit for the Derby. The show horse, however, must be produced at a peak for several months, during which time the owner must concentrate constantly on every detail of stable management and schooling.

A show horse should look a picture of health. He must not be fat or unfit; his coat should be like silk and he should be trimmed to an inch. Getting the coat right is a combination of correct feeding, good strapping and exercise. Feeding is, of course, a subject on its own, but from the showing point of view incline towards increasing the bulk foods and, particularly with the hotter type of show horse, decrease the high-energy intake. To help gain that lovely roundness, without 'flab', and high-gloss coat, small amounts of boiled linseed fed two or three times a week in the night feed are beneficial, as is a proprietary additive containing seaweed. Very often it is better to feed a sensitive horse four times a day, as a late-night feed will encourage him to eat up and digest his feed in peace.

The show horse should have a good top line, which comes with correct schooling. It is important to remember that he is well shod, and here it pays to find the best blacksmith you can—a farrier can make or mar your horse's feet and way of going.

Make sure that your tack fits properly, is correct for the class and that you know what time your classes are. Do not have a panic on show day—allow time for traffic hold-ups, punctures and all the other infuriating setbacks that can happen. It is better to arrive early and have time for a coffee!

Practice will tell you how long your horse needs for working-in prior to the class, remembering that manners in both horse and rider are very important. The modern show horse has a great deal to put up with (hot air balloons, motor-cycle displays, banners, screaming children, to give but a few examples) and it would be sensible to take a novice horse to a few shows without competing at first to get him to accept the sights and sounds, especially if he is of a highly strung disposition.

Schooling

The work you have done at home will be reflected in how your horse goes in the ring. Many people resort to a great variety of artificial aids to put their horses in an outline, but this is a poor substitute for correct work; one could do no better than study dressage, for though a show horse is not a dressage horse, the basic schooling principles are the same. The horse should go evenly on both reins, going forward and accepting the rider's aids, and carrying himself with elegance. Nothing is worse for a judge than to get on a mouthy, crooked horse, which appears to have no semblance of education. If a horse's conformation is good, he should 'ride you

right', but it is sad how often this is not the case because of poor schooling. It is also worth remembering that your own riding should be of a standard to help the horse. All too often one sees the case of a high-class horse gradually getting worse and worse with an inefficient rider. Take time and trouble to get help from a professional and constantly keep trying to improve yourself as well as your horse. In other words, if you set out with a goal, you should leave no stone unturned in your efforts to achieve it.

Your horse, even in a ridden class, will have to be shown in-hand for conformation assessment, and if you show in-hand alone—for example, youngstock classes—your in-hand presentation is as important as the ridden. The horse must stand correctly, and when being led should walk freely forward alongside the leader. When you turn to trot back, always

Overleaf Assurance—the perfect champion show hunter. With good conformation, strength and stamina—a classic horse for a day's hunting.

190

Right *Your horse's mane should always be plaited for competition unless the horse is a native breed or Arab. The mane should be kept pulled (remove a few hairs at a time from the underside of the mane by twisting them round the mane comb) to keep it manageable.*

Left top *Standing a horse correctly for the judge to assess conformation in hand.*

Left *The bridle of the show riding horse can be more 'showy' than the hunter's but the rider's dress should be similar. Gloves should always be worn.*

Far left *The show hunter's tack should be workmanlike and unfussy, and the rider should wear 'ratcatcher': tweed jacket, tie and hard hat.*

Dampen the mane on the offside to make it manageable.

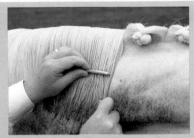

Decide how many plaits you wish to make and divide the mane with a mane comb.

A short neck can look longer with many small plaits, but this would not suit a large hunter.

Divide each, equal-sized section into three.

The plaits should start at the top of the crest and be pulled tight to give a neat, firm appearance.

Secure the end of the plait with double thread that is the same colour as your horse's mane.

Roll each plait up evenly to match the others.

Push a blunt-ended plaiting needle through from the bottom of the plait to the top and back again until the plait is secure.

This method of sewing up the plait makes it more difficult to undo, but gives, by far, the neatest finish.

turn the horse away from you so that he turns tidily and on his hocks, and trot in a straight line towards the judge, and on past him or her. Trot steadily, never dragging the horse. This is one of those important details that should be well practised at home, and if your horse is well presented it could make a great deal of difference to his final placing.

Ring procedure

When you enter the ring for your class, make sure you give yourself plenty of room—don't follow on the heels of the horse in front. Get your horse walking firmly forward, so that the judge's first impression is a pleasant one. If you find the class is bunching up, look for a gap and ride quietly across to it. Do not ride slap in front of the judge's toes, where he or she cannot get a good view of your horse. Once a trot is signalled, settle your horse

into an even rhythm, keeping well to the outside of the ring. Do not do a running extended trot; this is very ugly and does not show your horse off at all. The same applies to the canter, though if in a hunter class, the canter should be a little freer than in a hack or riding horse class. If you are required to gallop, do not drop your reins and go flat out—it is dangerous and makes the horse look scrambled and hurried. Ask your horse to lengthen his stride going into his bridle so that he gives the impression of scope and power.

When the ring steward has signalled the class to walk, come in from the outside of the ring and walk closer to the centre. The judge will then begin to select the exhibits he or she prefers. When the steward calls you in, form up quietly with the other competitors, leaving enough room between yourselves to avoid kicking each other.

At this stage in hunter classes, the judge will ride the horses, but in hack, pony and riding horse classes, each rider will be required to give an individual show (this is dealt with in the next section). In all adult classes where the judge may ride the horses, do make sure that your stirrup leathers can be made short or long enough, and that your stirrup irons are big enough. Not everybody is a uniform size! Some judges like a leg-up, which the steward usually gives, but your homework should have ensured that whether the judge is legged up or mounts by stirrup, your horse will stand quietly while he does so. After the judge has ridden the horse, thank him or her and take the horse back into the line. At this point you should have an assistant, who should be tidily dressed, preferably in jacket, hat, breeches and boots, or slacks, to come into the ring to remove the saddle and tidy the horse up to prepare for the in-hand section of the class. The assistant should have a grooming kit and, if the weather is chilly, a rug of some sort, because if it is a large class the horse may have to stand for some time.

When it is your turn to go out for conformation assessment, make sure you do not keep the judge waiting and that, on returning to the line, your assistant helps you re-saddle the horse. When this has been done, mount again and stand in the line. After all the horses have been out in-hand, the steward will ask the class to walk round in a small circle while the judge decides on the final placings.

This final walk round is important—you should try to keep your horse walking out well on the bit so that he presents the best possible picture. The steward will then, on the judge's direction, call in the prizewinners. If you are lucky enough to be one of them, go into your place and wait to receive your rosette, after which you will canter round the ring in a lap of honour.

There will be shows in which you feel that you could have come higher in the placings, but though we all like to win, it is very bad manners to make a scene. Remember that the judge has his or her own opinion and if your horse did not find favour on this occasion, there is always another day. You must also be totally ruthless with yourself in examining how your horse went and how you presented him. If you feel that things could have been better, then it is back to the drawing board so that next time there is a marked improvement.

Individual show

This is a very good opportunity to show how well your horse can go, so that if you have been 'missed' in the preliminaries, you now have the judge's whole attention.

Wait until it is your turn, and come straight out of the line-up with the pattern of your show and where you are going to do it already decided. The great thing about this individual show is not to be

Brush the dock through with a water brush to dampen it.

Tightly grasp two pieces of the tail from either side at the top of the dock.

Introduce a few more hairs from each side to make the three strands of the first plait.

Continue the plait by bringing a few more hairs from the back of the tail each time.

Take only a little hair at a time or the plait will be too thick.

At the end of the dock sew the plait up in a loop to the point it left the dock and then sew the loop flat.

Above *Plaiting the tail.*

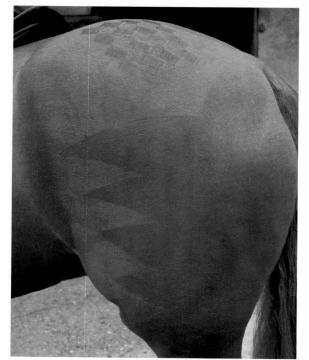

Left *Quarter markers are applied by brushing the coat in opposite directions with a damp cloth. A back may be given shark's teeth and chequers but a hunter should only ever display the former.*

Above A correctly turned out show pony and rider. A rider's turnout should in no way be theatrical. Hacks and ponies can wear very fine bridles while hunters and cobs should wear broader leather. All tack must fit properly and, if new, should be well worked-in and darkened before the show.

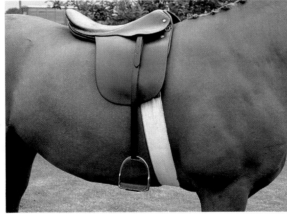

Right A straight-cut show saddle is preferable for all show classes as it displays the shoulder to best advantage. Dressage saddles are popular but sometimes their thick padding makes them ride up the horse's shoulder.

ual show' is the jumping test, and American show hunters are also required to jump a straightforward course of natural-type obstacles in a smooth, easy, forward-going style.

Tack and turnout

The way in which you and your horse are turned out is of prime importance. Your horse's mane should be plaited and his tail pulled or plaited, unless he is in a native pony or pure-bred Arab class. Practise your plaiting at home so that by the time show-day comes along, you know how long it will take and the plaits will be tidy. A well-pulled mane makes plaiting easier. The horse's whiskers and the edges of his ears should be trimmed, and his tail should be levelled at the bottom and not left too long.

You may put patterns on the horse's hindquarters, but never put chequer-board patterns on a hunter.

Just before you go into the ring, wipe a little baby oil around the horse's eyes and muzzle to give him a fresh appearance and of course oil his feet.

The tack the horse wears must fit correctly and be correct for the class. It is permissible in working hunter, working pony and four-year-old hunter classes to wear a snaffle bridle and in novice show pony classes it is compulsory to wear a snaffle. In other ridden classes, a double bridle is correct, though some people use pelham bridles if their horses go better in them. Coloured browbands are out for hunters, but acceptable for hacks, show ponies and riding horses.

The fitting of a bridle is very important—an incorrect fitting will make the horse go badly, so get expert advice if you are in doubt, and of course make sure your horse has been used to wearing the bridle at home. Hunters and cobs should wear broader leather, but hacks and ponies can wear very fine bridles.

Finding the right saddle may mean an exhausting search round all the local saddlery shops, but it will be worth it. The saddle must obviously fit the horse and be comfortable to ride in, but it should also be of the straight cut panel variety to show off the horse's shoulder. Dressage-type saddles are becoming popular, but they can be too thickly padded, which often makes them ride up the horse's shoulder, so take this into account. A leather girth is best because it is the least obtrusive.

Working hunters may be shown in any tack which would be used for hunting, except boots, but of course you would do yourself no great service by appearing in a lot of 'gadgetry'.

In conclusion

Showing is enormous fun but, like all other aspects of equestrianism, very hard work. However, if you have a nice horse or pony, the motto is—have a go!

too ambitious and attempt movements for which the horse is not ready. Each society issues a directive about what the show can contain, and it is up to you to string it together to make the movements you incorporate look smooth and show your horse to his best advantage; for instance, don't attempt to rein back unless your horse does it well, and above all, do not make your show long-winded—you will only irritate the judge and the other competitors. When you have finished your show, salute the judge and return to your place. Basically, try to show the horse at trot and canter with changes of rein and, in pony and riding horse classes, a gallop on to finish. A lot of horses begin to anticipate in an individual show, so it is wise to change the pattern of your show at times to prevent this happening.

In working hunter classes in Britain the 'individ-

12

SIDE SADDLE

Side saddle riding is wonderfully evocative of the elegance and femininity of a former, more leisurely era. But today it is far from an anachronism and increasing numbers of riders are discovering that 'side' is not only safe, but is really quite easy. In this sport, however, put saddle before horse: a prerequisite of learning to ride side is to find the right saddle, for ill-fitting saddles are the cause of most bad side saddle riding and most sore backs in side saddle horses.

Choosing your saddle
Although new saddles are available again, it is not safe to assume that they are designed correctly, and the advice of an expert should always be sought before buying. The classic makes of side saddle, produced between 1900 and 1940, are Owen, Mayhew, Whippy, Champion and Wilton, and Martin and Martin; and all of these, as well as having a seat shaped to place the rider's weight onto her right seatbone, will have a quick-release device at the top of the stirrup leather. Whilst the rider who falls off to the near-side of a side saddle can easily release her foot from the stirrup, a fall to the off-side could result in a trapped foot and a dragged rider if the stirrup does not have this safety device. Before this was invented, various types of 'safety' irons were used, but none of them was really satisfactory.

The saddle itself should have a seat that is level from front to back and almost level from side to side. It is not easy to sit completely straight on a dip-seated saddle, and very difficult to rise to the trot or jump in such a saddle as the rider's weight has to be lifted up as well as forward. A dip-seated saddle may be tolerable if the dip fits the rider's bottom, but otherwise the rider must sit where the saddle places her rather than where she chooses. Most people nowadays prefer a doeskin seat to a plain leather seat, as this is less slippery.

Ideally, the pommels should be large, and wider at the base than at the top, unless you have short thighs, in which case you may be more comfortable with straight pommels. The top pommel, or 'fixed' head, is the most important to consider when buying a saddle, for it cannot be moved or altered, as can the lower pommel or 'leaping' head. The latter is screwed into the saddle, and an extra screw hole is often provided to offer a choice of pommel positions. The pommel itself is made of almost rigid leather mounted on a metal bar, and this bar can be bent or straightened if necessary to fit the curvature of your thigh.

The fixed head is built into the saddle, as part of the tree, and any alterations to it are a major operation requiring complete stripping of the saddle, and are therefore very expensive. It is the position of this pommel that determines whether or not you can sit correctly in the centre of the saddle with your thigh bone parallel to the horse's spine. The amount of flesh round the knee varies from rider to rider; if you are well-fleshed, you will need a pommel further to the left than a rider with little flesh. The latter trying to ride on a saddle made for the former will have a thigh that aims out over the horse's left shoulder and vice versa, and neither will be able to sit straight. Minor adjustments can be made for the thinner lady by placing a pad between pommel and knee, but this is not satisfactory in the long term.

The length of saddle you need is related to the length of your thigh, from the back of the knee to the back of the seat. Ideally there should be 2 inches (5 cm) of saddle behind the rider, but a saddle that is too long is better than one that is too short, as the rider's weight on the very back of the short saddle will push it down onto the horse's back and cause sores.

Side saddles are measured across the centre of the seat, from the back of the cut-out to the back of the cantle. *As a rough guide*, a rider 5 feet 4 inches (1.6 m) tall will need a 16-inch (40 cm) saddle, and every additional 3 inches (7.5 cm) of rider needs another 1 inch (2.5 cm) of saddle.

It is possible to hire side saddles by the season, and this may be worth considering if you are likely to grow substantially. However, all the above considerations of fit should be taken into account,

194

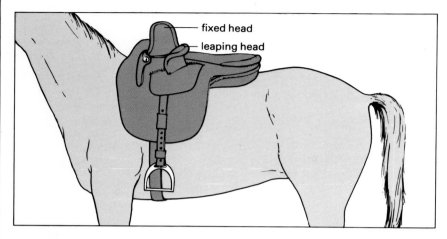

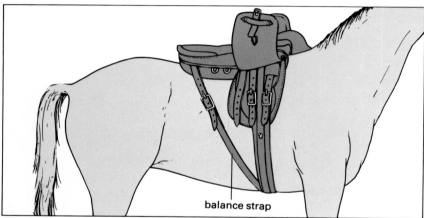

balance strap

Above A well-designed side saddle with a level seat and pommels wider at the base than the top. The upper pommel, or fixed head, which supports the right leg, is built into the tree of the saddle but the lower pommel, or leaping head, is moveable. The balance strap on the off side keeps the saddle firmly in position so that it does not give the horse a sore back.

Overleaf Jumping side saddle is safe and secure providing the rider makes no attempt to grip with the left leg. For fences under 4 ft (1.2 m) high the seat should be kept firmly in the saddle, the rider pivoting forwards from the hips.

and it is unwise to undertake leasing a saddle until the right one has been found. Side saddles hold their value well, and a good saddle may be regarded as an investment more likely to increase in value than decrease.

Having found a saddle that fits you, it will probably be necessary to have the stuffing adjusted so that it also fits your horse. Often more is needed on the nearside, both at the back and under the tree point in front, where this has been compressed by use. For this reason it is not wise habitually to mount from the ground, as this tends to pull the saddle over to the left.

Without a rider, a correctly fitted side saddle should sit with its left side higher than the right, and the gullet should be just to the right of the horse's spine. It will then settle centrally when the rider is on board. Ill-fitting saddles, especially those that tilt to the left, make it very difficult for the rider to sit properly, and the constant drag of weight will scald the horse's back and make it sore.

A suitable mount

Almost any horse can carry a side saddle, but those with wide withers and square shoulders may make saddle fitting difficult. For this reason, some Ara-

bian types may not be suitable, for their shoulder/wither shape encourages the saddle to roll and slip back. Horses under five should not be asked to carry a side saddle for any length of time, as it will put an excessive strain on their immature back muscles.

Any horse will carry a side saddle—but those with straight shoulders or high knee action may give a rather jolting ride. Lack of schooling, with examples of idleness such as going on the forehand, or dropping the inside shoulder on corners, will feel worse side than stride, but any well-schooled, well-balanced horse will soon adjust to the lack of a right leg. Many horses actually go better for their owner under a side saddle; which should tell you something about your seat and hands when you are astride.

Riding side

The main things to remember about riding side saddle are that you must sit straight and centrally, and that your seat balance and security is based on your right leg and its purchase against the fixed head and the horse's shoulder; not your left leg gripping up under the leaping head.

The best way to start off sitting properly is to mount stride and settle centrally this way before lifting your right leg over and putting it round the fixed head. You will find that if you first place your right hand on the back of the saddle, and keep it there, this will prevent the right side of your body following the leg round.

The left leg should be in a normal riding position, with the stirrup adjusted so that your left thigh is about an inch away from the leaping head. If the stirrup is too long, your leg could escape out from under the pommel in an emergency. If it's too short, forcing the leg up tight under the pommel, the left hip will be pushed back and your body will twist. Even with a correctly adjusted stirrup, there is a temptation to grip up under the leaping head, but this is wrong.

When the right leg is round the fixed head, there should be one or two fingers' space between it and the back of the knee. The thigh muscle should be pulled to the inside, so that a flat surface rests on the saddle and the knee is really snug against the pommel. The lower leg should hang naturally, with the toes down a little, and the calf pressed against the flap of the saddle. This is the purchase referred to above. The toes should not be turned up, nor the calf muscles tensed, or the knee will be pushed up off the top of the saddle away from the pommel. Think of 'pulling the knee down with the toes' to describe the correct position. The most secure purchase is obtained when the toes are turned inwards slightly towards the saddle.

When your legs are in position, consider your hips and shoulders. It is tempting to drop your left hip, especially if the saddle is not quite level, but

The rider mounts astride and positions herself comfortably.

With one hand on the cantle, she maintains a central position.

The right leg is swung over the upper pommel, or fixed head.

The rider keeps her central position, the right hand preventing her body twisting.

She makes sure there are one or two finger's space between her knee and the fixed head.

With her right leg secured around the fixed head, the rider again checks her seat.

A good side saddle seat. The rider sits straight and centrally, with her left leg in normal riding position, toes turned slightly inwards, and right leg secured.

Above *Mounting side saddle.*

Bottom left *The saddle, not quite level, has caused the rider's left hip to drop.*

Bottom right *If the rider slouches in the saddle the weight is unevenly distributed.*

this causes your whole body to twist and, apart from making you sit sideways, also loosens the right knee from the pommel. No amount of trying to 'put the right shoulder back' will correct this fault, nor will imagining that there is a tintack under the left buttock, as suggested by so many instructors. A normal reaction to this command is to *lift* that buttock, which tilts the hips and pushes the upper body out to the left. Since the cause of the problem was allowing the left hip to drop back, all that is needed to correct it is to bring that hip forward.

A useful exercise which will help to straighten your body, and give you the correct feel, is to face front and lift your right hand, palm up, on a straight arm until it is level with your shoulders. Then, without moving your legs, swing your arm out to the right, with your body and head, until it is at right angles to the horse. Drop your arm and relax your body. You will feel a slight pull across the front of the ribcage, and the right leg should feel very secure against the pommel.

In the correct position you should sit up straight, but relaxed. The back should not be arched rigidly, but pulled up to its full height, with the shoulders back and the lower ribcage pushed a little forward.

You may be tempted to drop the right shoulder, or lean out to the right, neither of which is correct, and someone on the ground behind can check.

If no helper is available, it is possible to check your position from the saddle. Without moving your head, cast your eyes down and visualize a line running down your breastbone from top to bottom. When this line gets to the navel, it should be turned through a right angle, and taken out in front. If you are sitting correctly, and the horse is standing straight, this line should go up the centre of the horse's neck and between his ears. If it goes off to the left, the left hip has probably been dropped, and if it goes to the left and up in the air, then you are leaning out to the right. All you need do is adjust your body position until the line is in the correct place, and keep it there.

The hands should be level, and carried in a position that maintains the bit/rein angle the horse is accustomed to. On an inexperienced horse, this will mean somewhere close to the right knee, but on a well-schooled, well-balanced horse, they may be carried in the lap. Unless using the right rein in opposition behind the wither as an aid to lateral movements, the right hand should not be dropped back and down below your seat, especially if this position is accompanied by a rigid right arm.

A whip is usually carried in lieu of the right leg, and this should be long enough to reach where the leg would be, and rigid enough to give the same feel to the horse as a leg would. Thin dressage whips are often too whippy. The whip is used either by flicking it to produce light taps, or by turning the hand to the outside so that the far end of the whip presses on the horse's side.

Training the horse

You will find it better to restrict the first few practice sessions to the right rein, where it is easier to balance the horse on corners and to maintain your body position correctly. You may find it necessary to use the whip in conjunction with your left leg in the early stages, but most horses soon adapt and accept leg aids from the left only. As you become more experienced, you will learn to use your hips and body weight for fine control of your horse.

Many novice riders experience difficulty in obtaining right canter, but this is usually because they drop their left hip when putting their left leg back for the aid. This weight shift tends to bend the horse to the left, especially as the left hand is brought back at the same time, which is inevitable when the left side of the body comes back. The trick of obtaining canter side saddle, on either rein, is to lift the inside hand to obtain some bend, turn the shoulders in the same direction as the desired bend, and nudge inwards with the inside seatbone when the horse is on the trot diagonal of inside hind, outside fore. This nudge has the effect of telling him to tread well on his inside hindleg and to start the first phase of canter with his outside hindleg.

Your back and waist should be kept supple at all paces, but especially at the trot, or you will bump badly. Posting the trot is obtained by shifting the weight from back to front of the right thigh, and back again. If you try to rise by pushing off from the stirrup, you will rise out to the right side and the saddle will shift and rub.

It is necessary, for both elegance and security of seat, to sit up very straight. A little slump may not be noticed astride, but looks awful on a side saddle, and anything that lifts the seat lessens security.

Keeping hold of the horse's head, the rider releases the elastic loop, which secures the apron to the boot.

The apron is tossed over the rider's left arm to keep it out of the way.

Launch yourself carefully off the saddle.

When jumping down the rider holds the fixed head for security.

Secure jumping

Jumping side saddle, like the whole seat, is dependent on the right thigh, not the left leg and leaping head. Gripping with the left leg when jumping is a major cause of falls, for it pushes the left hip back, which pushes the right hip forward and loosens the right knee, and the end result is that the rider swivels right out of the saddle, landing on her back.

The seat should be kept on the saddle at heights under 4 feet (120 cm), and until the correct feel of jumping side is firmly fixed in the subconscious. All that is necessary is to pivot forward from the hips, and to allow your hands to precede you. You should not drop your hands down the horse's shoulder, nor attempt to get forward out of the saddle; you will only receive a nasty jolt when you land, with a resulting whip-lash effect on your neck.

Correct habit

The correct attire for riding side saddle is called a habit, and this consists of a jacket and matching apron. The jacket will be more or less like an ordinary riding jacket, but cut away in front so that it lies neatly across the thighs. The apron fits over the right thigh and knee, and tucks behind the right calf, where it is secured by an elastic loop over the right toe. It buttons up on the left hip, and should be long enough to cover the right foot at all times,

Above *The correct way to dismount.*

Right *The correct habit for adult and child; a picture of elegance.*

The right leg is swung over the fixed head while the left leg quits the stirrup.

With both legs free, but still a hand on the reins, the rider jumps to the ground.

...ce on the ground the rider can let the ...ron fall.

After dismounting the rider has a friendly word with her horse.

and should hang parallel to the ground. For showing, top boots and a waistcoat should be worn, and ideally matching breeches. A blunt spur should be worn on the left heel.

Local rules on colours and headgear should be checked before competing, but it is correct at all times to wear a veil with top hat or bowler; a top hat should only be worn with a black habit. Traditionally, black gloves should not be worn, and tan or beige are preferred.

In costume classes, historical accuracy should be maintained, and flamboyant tea-gowns or trailing draperies avoided. For western classes, a non-flowing skirt or gaucho pants may be worn. The latter may need to be longer in the right leg than the left, so that the tops of both boots are covered. With either alternative, a long-sleeved shirt should be worn, with necktie, kerchief or bolo tie, and a western hat.

Try side

With the sole exception of classes run under American Quarter Horse Association rules, side saddles may be used at any time in the show-ring, for hunting, dressage or riding club activities.

Other hunt followers will be delighted to see a lady side saddle, and many ladies prefer to hunt side from a safety angle. Although it is by no means impossible to fall off a side saddle, especially in the early stages of learning, there are many situations in the hunting field where an astride rider falls off and a side rider stays on, even if the horse has fallen and is grovelling in the mud on his knees. It is also quite easy to step off a side saddle if the horse is falling to the left, so there is little danger of being crushed. The only hazard attached to hunting side is that of low branches in woody country.

Many shows now offer side saddle equitation classes, and the Side Saddle Association and International Side Saddle Organization hold special 'side saddle only' shows. There is no reason why a horse should not be shown side saddle in any class, but it may be advisable to offer a cross saddle for the judge to ride on.

Dressage tests may be ridden side saddle at all levels providing the organizers are forewarned. In some early tests, the rider must rise to the trot where the test requires. A whip of not more than 3 feet (1 m) in length may be carried at all levels.

Considering that pessimists have been announcing the demise of the side saddle ever since the turn of the century, it is most gratifying to all enthusiastic 'Amazones' that circumstances have proved them wrong. Far from disappearing, side saddle is becoming even more popular as a new generation of riders discovers the merits and delights of this elegant method of riding. It is also a very invigorating and exciting way to ride a horse and certainly well worth trying once you have achieved a good level of horsemanship.

13

DRIVING

For the growing number of enthusiasts who prefer to sit behind their horses rather than on top of them, a variety of sporting and recreational activities has gradually replaced practical horse transport over the last sixty years. For a generation after World War I, the art of the coachman was almost forgotten in Western countries: the motor-car had come to stay. But aficionados remained and the past decade has seen an enormous upsurge of interest, especially in the newest form of driving, the competitive performance trials held under international FEI rules which take place throughout Europe.

Requirements and cost
Most breeds of horses and ponies go well in harness, the possible exceptions being those with a preponderance of Arab or Thoroughbred blood which are bred to gallop rather than trot. Driving horses do not need exceptional physical ability, like racehorses or show jumpers, and good action coupled with calm, courageous temperament may be more important to them than perfect conformation. Welsh cobs, Welsh Mountain ponies and Hackneys dominate the British driving scene, with Shetlands holding their own remarkably well in trials for teams and pairs. Cleveland Bay influence is also still in welcome evidence, particularly in the Royal Mews. Hungarian horses and Dutch Gelderlanders are popular choices for the coachman, while Polish horses have the attributes to be popular if their owners were better salesmen! Lipizzaners are making their mark in Britain, America and South Africa, but Kladrubers from Czechoslovakia are the most impressive carriage horses in the world, and if the Russians ever enter the international driving scene, their Orlov trotters will take some beating.

Several cobs and ponies have literally been saved from the knackers to win in the show-ring and in driving trials, and the potential buyer will find an open market here. Prices of horses in Britain are generally lower than in Western Europe or the United States, and there are variations everywhere,

but at the time of writing £800 will buy a harness horse with plenty of potential, and £500 a good driving pony. The rest is up to the driver and offers a fascinating and rewarding challenge.

The smartest turnout for a driving horse is reckoned to be a painted gig or two-wheeled dog cart, and black patent leather brass-mounted harness. You don't have to have this to win a show class, but it may help, and the harness will cost up to £900 in Britain; British harness is considered to be the best produced and prices are generally slightly higher than in other countries. There is some fine old harness still about, whose stitching and brass furnishings may be better than those of today, but if it is more than fifty years old it will not have kept its condition unless it has been very well looked after. Brown harness is rather cheaper and perfectly suitable with a varnished cart for a more country turnout.

There is now plenty of webbing harness being made, with traces of the same material as car safety belts for added strength. This is sold for about £50 per single set and does very well for exercising and the marathon competition in driving trials; it is easier to clean and repair than leather harness. Harness is adjustable within a height range of about one hand (10 cm), and this includes a breast collar when one is used. A full neck collar, with its essential hames, must fit the horse that wears it to within half an inch (1.2 cm) and is much more expensive than a breast collar. It is more formal and traditional with English harness, but only more efficient when a heavy load is involved. Pair harness and tandem harness is nearly double the price of single harness.

A two-wheeled cart for a single pony will cost upwards of £200. It will fit a pony 2 inches (5 cm) shorter or taller than the exact height for which it is built, and the cost will increase by about £50 per hand for bigger sizes. £200 will just about buy you an exercise cart for everyday driving around your own home, but you wouldn't want to rely on it for driving trials and it will do you no credit in the show-ring. You must consider a vehicle costing

£500 upwards for driving trials and a price of at least £800 to complement a potential winner in a private driving class at an established show. British prices again tend to be higher, especially for the best-made vehicles.

A carriage-builder will usually build a vehicle to your own specifications, or to a traditional design, and many of them will expertly restore an old one, but the process, and particularly the painting, is very time-consuming. It is consequently not cheap, starting at around £500, but restoration in America tends to be more expensive. Governess carts are good for their original purpose, the transport of small children, but not much favoured for competitive activities in which the driver has best control on a fairly high seat facing forward. Dog carts and Ralli cars are excellent in this respect, but a light gig may just have the edge over these for displaying a horse to his best advantage in the show-ring, particularly as his action may suffer if he has to pull much more than his own weight on soft grass. Any of these will suit a tandem, but you need a four-wheeled carriage for a pair or team and this will inevitably cost more than a two-wheeled vehicle; usually more than half as much again. Nearly all serious driving trials' competitors now use modern steel and aluminium carriages, at least for the marathon, and pay up to £5,000 for them. Some very fine old carriages have changed hands for more than this, the record being £15,000 for a park drag, but these will last for 100 years. Indeed, many of them have already done so.

In budgeting for capital cost you should add £50–£100 for the lamps, whip and other appointments to complete a single turnout. As a very general estimate, a complete single turnout of decent standard will cost between £1,500 and £2,500. You must add 50 per cent to this for a tandem, almost double it for a pair, and be prepared to quadruple it for a four-in-hand. These figures do not include a reserve horse or a spare vehicle or harness, or the motor transport for travelling to shows and events, and this last item is likely to be by far your biggest expense.

Overleaf *Combined driving is the most spectacular of equestrian sports.*

Above *To the rider, driven dressage tests may appear simple, but the extra difficulty of encouraging four horses to work and move together with precision more than compensates for this. Unlike ridden dressage, the voice may be used and, if discreet, the groom may remind the driver of the next movement!*

FEI Driving Trials

International driving trials are perhaps the most exciting modern development in harness sports. Their universal appeal has spread from Europe to Britain and is now emerging in America, Australia and South Africa. They are based on ridden horse trials and follow their format closely, consisting of three separate competitions which together test all the qualities and abilities of horse and driver. Currently, only four-in-hands compete in international trials, but national rules cover classes for singles, pairs and tandems as well, and these are sub-divided for horses and ponies at the international height limit of 14.2hh (145 cm). A further sub-division for pony heights at 12.2hh (125 cm) is sometimes recognized nationally but not internationally

Competition A in driving trials contains two elements: A(i) Presentation and A(ii) Dressage. Presentation is an assessment of appearance at the halt in which bonus marks are awarded in accordance with the dressage scale for five aspects of the whole turnout: driver and groom(s), horse(s), harness, vehicle and spare equipment, and general impression. The average of the marks of the three judges is subtracted from the possible maximum of fifty and shown as penalty points for each competitor. Quality alone is not the deciding factor and the judges concentrate on condition, cleanliness, good fitting and suitability. A good rule for competitors is 'keep it simple'. The more you add by way of embellishments, the more there is to arouse criticism.

A driven dressage test is similar to a ridden one but is judged by five or three judges in a bigger arena. All current tests consist of eleven movements and four assessments of general impression. The average of the judges' marks is subtracted from the possible of 150 and recorded as penalty points.

Competition B, the marathon, is the equivalent of the speed and endurance phase of a ridden three-day event and is of course the most exciting. There are three trot and two walk sections in a full three-day event, of which the last, at the trot, includes between five and eight obstacles. National rules allow these to be reduced to two trots and one walk for a two-day event and to one trot only, including the obstacles, for a one-day event. The total length of the course for an international event must be between 23–27 km (14.3–16.8 miles), but this may be reduced to a minimum of 15 km (9.3 miles) or 6 km (3.75 miles), according to the number of sections. Penalty points are awarded in accordance with time taken over each section and through each obstacle, and for a variety of specified faults and errors.

Competition C, the obstacle competition, replaces the show jumping phase of ridden trials on the third day. Competitors have to negotiate a tortuous course through closely spaced traffic cones from which balls fall if touched, at a cost of ten penalties each time. Any pace is permitted, but a speed of 200 metres (220 yards) a minute or faster is required, and exceeding the time allowed, which is based on this speed, incurs a further penalty.

The competitor with the lowest total penalty points in his class for all three competitions wins the whole event. A driver who is eliminated or retires in any of the three competitions is given the score of the lowest-placed competitor in that competition, plus 25 per cent, and may continue to complete the remaining sections. He may not subsequently be classified above any competitor who has completed the whole event, or win a prize for it, but he may be placed in any of the three individual competitions and win any special prizes awarded for them. The scores recorded for national teams are those of the best two of the three members in each of the competitions, except that the score of a member who is eliminated or retires may not be credited to his team for any competition.

Driven dressage

Driven dressage is performed in an arena of 100 metres × 40 metres (110 yards × 44 yards), but this may be reduced to 60 metres × 40 metres (65 yards × 44 yards) for single turnouts and pony pairs. The markers are the same as for ridden dressage and when there are five judges they sit at C, F, H, K, and M while a jury of three is placed at C, B and E.

There are three different tests in the current international rule book: two graded advanced and one elementary, each timed for ten minutes in the big arena and eight in the smaller one. There is also a five-minute novice test in the national rule book, and the British Driving Society has a new dressage test of its own.

Movements are simpler than those in comparable ridden tests since, obviously, there can be no lateral work and circles are not less than 20 metres (20 yards) in diameter. There are fewer paces with less variations of them, since only the ordinary walk, the working, collected and extended trots, and the rein-back are demanded. Relative simplicity in these respects helps to compensate for the extra difficulty of getting two or four horses all working and moving together with precision. Unlike ridden dressage, the voice is recognized and permitted as an aid in place of the rider's legs, as is the whip.

If they are to learn instant obedience and develop correct paces, harness horses must be well ridden as well as driven, and small ponies may be long-reined instead by experienced trainers. Remembering the test is often the biggest worry of novice drivers, but this is easier for them than for riders, since driven tests always form four or five obvious patterns, which follow a logical design once they are started and, although grooms are not allowed to give obvious indications, they may remind their drivers discreetly of the sequence of the patterns and the paces required. As with riding horses, it is a mistake to practise complete tests too often, if at all, particularly in the same arena, as this results in boredom and anticipation. It is more refreshing for driver and horses to practise paces and individual movements during outings around the countryside.

The marathon
The marathon starts with Section A, which is a trot over roads and tracks for a maximum distance of 10 km (6.25 miles) at an average speed of 15 km/h (9.3 mph). Section B follows immediately and is a walk of up to 1200 metres (1300 yards) at 7 km/h (4.3 mph) usually 6 km/h (3.75 mph) for ponies. There is a ten-minute compulsory halt and a veterinary check before Section C, which is a fast trot of about 5 km (3 miles) at a speed of up to 20 km/h (12.5 mph). Walk Section D follows to the same specifications as B, and is followed by a second ten-minute halt and check. The last section, E, is over a maximum distance of 10 km (6.25 miles) at an average speed of 15 km/h (9.3 mph), including between five and eight specially built and marked obstacles. Competitors are penalized at the rate of one penalty point for every five seconds by which they exceed the time allowed in any section, and similarly for finishing more than two minutes early in Sections A or E, or more than one minute early in Section C. Penalty points on the same scale are awarded for the total time taken through each obstacle in Section E. A referee travels with each driver except those of singles and pony pairs, which are refereed from the ground, and he records any errors of course, breaks from the prescribed pace which last more than five seconds, and other faults or rule infringements. The obstacles are judged on time and a scale of specific obstacle penalties. A competitor who does not negotiate an obstacle in five minutes is eliminated.

The marathon provides a rugged, exacting test, not much related to elegant park driving, and horses should be hunting fit for it, having been in regular work for at least six weeks. They may be able to complete the course when tired, but will be less manoeuvrable in such a state.

It is a shame to risk fine old carriages on a marathon and the rules permit a second special vehicle to be used which should be, as far as possible, of steel construction, with disc brakes for a four-wheeler. International rules require horse team vehicles to weigh not less than 600 kg (1320 lb) and those for pony teams not less than 300 kg (660 lb). Minimum weights are likely to be specified in future for horse and pony pair vehicles at 350 kg (770 lb) and 225 kg (495 lb) respectively, and for horse and pony singles carts at 150 kg (330 lb) and 100 kg (220 lb). The last three of these

weights are less than those of most traditional carriages, so special construction is desirable. It is a considerable handicap for a horse to have to pull more than his own weight in hilly deep going. Refer to the Eventing chapter for relevant remarks on roads and tracks, course walking and so on, but make sure also that your groom(s) knows the course as well as you do for driving trials.

The obstacle competition
Competition C, the obstacle competition, is primarily intended to determine whether horses are still fit, obedient and responsive after the rigours of the marathon. It is also, however, a precise test of skilful and accurate driving, which is more demanding than it appears to spectators. The course consists mainly of 3-foot (1 m) plastic cones surmounted by balls, set in pairs from 1–2 feet (30–60 cm) wider apart from each other than the track width of the vehicle concerned. A water obstacle, a low wooden bridge and elements of show jumps may also be included. Ten penalty points are awarded for each ball or other element dislodged, and circling before an obstacle, stopping or reining back, or dismounting a groom is treated as a refusal with cumulative penalties as in show jumping. A time allowance for the course is calculated for a speed which may be varied between 220 and 275 yards (200 and 250 m) a minute, with half a penalty point awarded for every second by which it is exceeded. 'Drives-off' against the clock may take place to resolve equality and the competition can be judged on time alone by adding ten seconds instead of ten penalty points for each knock down. The tedium of resetting the cones between each round may be obviated in future by fixing a standard width bar to each vehicle. This will reduce measuring to a routine check of prize winners. Competitors are well advised to walk the course

Above *The marathon is a rugged, exacting test of horses and driver, whose 'steering' cannot afford to wander. Such big, solid obstacles as this one encountered in the 1982 World Championships are only argued with at the driver's peril.*

Left *The Duke of Edinburgh, an active and successful competitor, has done much to help the growth and popularity of the sport.*

several times with their grooms and plan a line to each obstacle, deciding where to turn to make a straight approach. The driver then aims his leaders(s) and leaves the steering up to the horse(s), keeping his eye on the pole head and aiming to steer this through the middle of each obstacle.

Conclusion

To people who regard horse-drawn carriages merely as outmoded forms of transport, those who drive them may seem to be old-fashioned eccentrics who have chosen an unusually expensive way of meeting their friends and astonishing their neighbours. A closer look from a horseman's point of view, however, will reveal a flourishing equestrian recreation in which many different breeds of horses and ponies can provide a wide range of activities for people of all ages and circumstances. Anyone who cares to take this closer look will appreciate the advantages, for example, of parents

being able to drive their children's ponies and keep them fit and sensible between school holidays. The same applies to summering hunters, which can be exercised most pleasantly with no weight on their backs.

In many other competitive horse sports, riders get too old too soon and retire just when their judgement and horsemanship have reached their peak. Driving makes fewer demands on agility, physical fitness and, some might say, nerve, so you can continue to take an active part for much longer. Harness horses do not have to jump or display absolute speed, and are only asked for simple paces and movements in dressage. Their experience and physical ability therefore count for less than the skill and judgement of their drivers. This is a significant factor in driving which makes it particularly attractive to people who aspire to the highest levels of competitive horsemanship but cannot afford all the time and money this usually costs.

INDEX

Page numbers in italics refer to illustrations.

Acknowledgements

The publishers wish to thank the following organizations and individuals for their kind permission to reproduce the pictures in this book:

All Sport Photographic 7; Animal Photography Ltd/Sally Anne Thompson 26 above inset, 40 above; Ardea London Ltd/Jean Paul Ferrero 2–3, 67 above, 92, 116–117, 120 above; Mike Busselle 172–173; Bruce Coleman Ltd 40 below; Kit Houghton 70–71, 126, 168 left; Bob Langrish 97, 169, 170, 171; Peter Roche 46–48, 50–55; Vision International/Elizabeth Weiland 62–63, 85 below, 101, 109 above, 164–165, 166, 167, 168 right

Special Photography
Kit Houghton 4, 5 centre and below, 12, 18–19, 20–23, 28, 29 above and below, 32–33, 38–39, 56 left, 57–61, 67 below left and right, 68, 69, 72–73, 75–78, 81, 82, 86–87, 92–93, 98, 100, 102–105, 107, 108, 109 below left, 110–115, 118–119, 120 below, 121, 122, 124, 125, 127, 132, 180–189, 190 above, 190 below right, 191, 192, 194–195, 197 above, 198–199; Ross Laney 109 below right, 158 above and below left, 159, 160 below right; 162, 190 below left, 193, 197 below left and right; Bob Langrish Endpapers, 5 top, 9 above left, 13 above, 16, 30–31, 34–37, 49, 74, 96, 99, 106, 123, 128, 133, 136, 137 inset left, 138–139, 140–157, 158 above and below right, 160 above left and right and below left and bottom, 161, 163, 174–179, 200–205; Chris Linton 26–27, 26 below inset; Sally Anne Thompson Half Title, 8–9, 9 above right, 10–11, 13 below, 24–25, 38 left, 41, 42–43, 56–57, 79, 80, 83, 84, 85 above, 88–91, 94–95, 129–131, 134, 135

The publishers would like to thank the following people for their kind help with photography:

Breeds:
Kentucky Horse Park, Lexington, USA
Ed Bryce

Side Saddle:
Headley Trace Stables
Anna Ahmad
Emma Burge
Tracey Gibbs
Judy Jolly
Carol McCulloch

Saddlecombe Stables
K.C. Colonna

Show Jumping:
Easton Grey House
Ian Adsetts
Peter Saunders
Rosalind Ward

Showing:
Shepperlands Copse Stables
Victoria Brooks
Lucinda McAlpine
Helen Pledger

Feed
E.J.O. Stone at Dalgety Spillers Agriculture Ltd.